EXTRATERRESTRIALS

EXTRATERRESTRIALS

WADE ROUSH

The MIT Press | Cambridge, Massachusetts | London, England

This book was set in Chaparral Pro by Toppan Best-set Premedia Limited. Printed and bound in the United States of America.

Library of Congress Cataloging-in-Publication Data

Names: Roush, Wade, author.
Title: Extraterrestrials / Wade Roush.
Other titles: MIT Press essential knowledge series.
Description: Cambridge, Massachusetts : The MIT Press, [2020] | Series: The MIT Press essential knowledge series | Includes bibliographical references and index.
Identifiers: LCCN 2019024257 | ISBN 9780262538435 (paperback) | ISBN 9780262357593 (ebook)
Subjects: LCSH: Search for Extraterrestrial Intelligence (Study group : U.S.) | Life on other planets. | Extraterrestrial beings.
Classification: LCC QB54 .R597 2020 | DDC 576.8/39—dc23
LC record available at https://lccn.loc.gov/2019024257

10   9   8   7   6   5   4   3   2   1

CONTENTS

SERIES FOREWORD

The MIT Press Essential Knowledge series offers accessible, concise, beautifully produced pocket-size books on topics of current interest. Written by leading thinkers, the books in this series deliver expert overviews of subjects that range from the cultural and the historical to the scientific and the technical.

In today's era of instant information gratification, we have ready access to opinions, rationalizations, and superficial descriptions. Much harder to come by is the foundational knowledge that informs a principled understanding of the world. Essential Knowledge books fill that need. Synthesizing specialized subject matter for nonspecialists and engaging critical topics through fundamentals, each of these compact volumes offers readers a point of access to complex ideas.

Bruce Tidor
Professor of Biological Engineering and Computer Science
Massachusetts Institute of Technology

We live in the middle of a great unfinished story: the tale of humanity's emergence as a planet-dominating but homebound species and our potential transformation into a spacefaring, multiplanetary one.

We know how the story began, as we left the treetops, spread across the continents, and developed language, farming, written culture, and eventually science.

We know where the story stands now. In the current century, we'll be busy confronting the multiple existential challenges we have created for ourselves. But if we can master the forces of global-scale politics and industrial-scale technology, own up to our responsibility to manage Earth's climate and ecosystems, and systematize our exploration of the other planets—all big *if*s, admittedly— then there is not much else to stop us from expanding outward into the galaxy.

What we don't know is which direction the story will go after that. There are only two possibilities. Either we will remain alone in our explorations, or we will find that we have company.

It's conceivable that we are the first intelligent beings in the galaxy ever to consider leaving their home planet. In this scenario, we'll learn that the rest of the Milky Way is home to microbes and little more. Everywhere we venture, we will find lonely and uninhabited spaces, waiting to be colonized—by us or by our artificially intelligent machines.

Or perhaps we will run into other people as we go, as we always have before. There might be a whole bustling, galaxy-spanning league of planets waiting to welcome us or perhaps a few scattered but nonetheless sociable civilizations.

If and when we do make contact with other cultures, they are likely to be far older than ours, and the interaction is likely to transform us in ways that are hard to imagine now. The astronomer Paul Davies, who chairs the Search for Extraterrestrial Intelligence Post-Detection Task Group, believes contact with extraterrestrials would have "a greater impact on humanity than the discoveries of Copernicus, Darwin and Einstein put together."[1]

But the suspense in this matter may last for a long time. If we are not the first galactic explorers, we might learn this tomorrow or a hundred years from now or a thousand. If the galaxy seems empty, we will have to live with our solitude forever, never knowing whether we are the first (or the last) or whether there are others, but they remain too remote for us to find.

To me, what's interesting is that we've grown up enough as a species to know how to *ask* the question "Are we alone?" but not enough to know how to answer it.

It would be reasonable to assume, for the moment, that we *are* alone. There's no physical evidence that aliens have visited our solar system.[2] We have been listening for radio signals of intelligent extraterrestrial origin for 60 years, and we have not heard a peep. The aliens might be hiding, or they might be too far away, or they might be communicating in ways we can't yet detect—but the most conservative guess right now is that they just aren't there.

But that's only a surmise. We don't know how life arises or exactly how many worlds are suitable for it. We don't know how often simple organisms evolve all the way to the point of sentience and tool making. We don't know how other cultures might try to communicate with us. So far we have chosen to search for their signals on radio and optical frequencies, but the use of those methods in particular, too, is just a guess. Our search may be doomed to futility, or we might just be getting started.

The question of the existence of extraterrestrials is not just one of the most persistent puzzles in science; it is the biggest blank in our own story arc as a species. And it is marvelously, tantalizingly unsettled.

That's what this book is about: the question itself, why it remains unanswered, and how scientists are trying to answer it.

The question has been nagging at my own mind for almost as long as I can remember. I was born in the late 1960s and devoured pop-culture artifacts such as Stanley Kubrick's *2001: A Space Odyssey* (1968) and Steven Spielberg's *Close Encounters of the Third Kind* (1977) and *E.T.: The Extra-Terrestrial* (1982). The idea that space aliens *might* exist and that it might be our destiny to meet them had long since become part of the zeitgeist.

As we will see in chapter 1, the concept goes back to the ancient Greeks and entered firmly into popular lore in the nineteenth and early twentieth centuries with the help of scientists such as the Mars-obsessed astronomer Percival Lowell and science-fiction writers such as H. G. Wells.

But, for me, it was the work of the renowned astronomer, science communicator, and TV personality Carl Sagan that brought the idea into sharp focus. Sagan was part of a small group of researchers who had been working since the early 1960s to make the search for extraterrestrial intelligence, or SETI, into a respectable scientific discipline. He wrote at length about extraterrestrial life in his book *The Cosmic Connection* (1973),[3] then led the production of the Voyager Interstellar Record. Encoded with audio and photos, the record went into deep space in 1977 aboard the Voyager 1 and Voyager 2 probes as a kind of aspirational message in a bottle. But perhaps most importantly, Sagan electrified my whole generation of budding science geeks with his public-television series *Cosmos: A Personal Voyage* (1980).

The penultimate episode of the show, "Encyclopaedia Galactica," was all about SETI. It explained that the question of extraterrestrial life was one that scientists, not just science fiction authors, could examine. By the time I reached Harvard College in the fall of 1985, I not only shared Sagan's optimism that extraterrestrials must exist but had become a full-fledged Sagan wannabe, choosing to major in astronomy and taking a work-study job at the Harvard-Smithsonian Center for Astrophysics, where he had once worked.

Imagine my excitement, then, when Sagan himself showed up on campus. He was there to participate in a symposium to christen the Megachannel Extraterrestrial Assay, a radio-frequency SETI project spearheaded by the Harvard physicist Paul Horowitz. I attended the symposium and approached Dr. Sagan afterward to share all my fanboy enthusiasm. He was as gracious and engaging as I could have hoped.

At the very same time, I was trying out for the news section of the weekly campus paper, the *Harvard Independent*. Thanks to the symposium, I had a story to pitch. The resulting piece about Horowitz's project, which had been made possible by a $100,000 gift from none other than Steven Spielberg ("I thought it was time that I got involved in a little science reality," Spielberg said at the symposium[4]) appeared in print the next week. It was my first published article of any kind, and it left me with a passion for writing about science and technology that has never let up.

Within a couple of years, my interest in journalism and the history of science had overtaken my interest in astronomy. I went on to get a PhD from MIT in the history and social study of science and technology, and I have spent my adult life working as a technology journalist for print, Internet, and audio outlets.

But I have long felt that my I owe my career, in some small way, to Spielberg, Horowitz, Sagan, and SETI. The funny thing is that after that first clip for the *Independent*, I never returned to writing about the subject—until the MIT Press invited me to contribute this book.

For me, then, this project has offered both a chance to complete an important life circle and a welcome opportunity to immerse myself, more than 30 years after that Harvard symposium, in historical and current thinking about the search for extraterrestrial life and alien civilizations.

Before going on to share what I have learned, I would like to thank Susan Buckley, my editor at the MIT Press on this project and a previous one, the hard-science-fiction anthology *Twelve Tomorrows* (2018). She solicited the proposal for this book and has been a creative and patient counselor.

In the fall of 2018, while I was doing the research for this book, it was also my honor to coteach a seminar on SETI for MIT's Experimental Study Group. My friend and coinstructor, the MIT astrophysicist Paola Rebusco, made that experience a joy and inspired me with her creative teaching ideas. Paola also took the time to review the manuscript for this book. The MIT first-years in the seminar—Annalisa Broski, Juliana Drozd, Raquel Garcia, Sarah Lincoln, Joshua Rodriguez, Elena Romashkova, and Talia Spitz—asked hard questions that also helped me sharpen all of the ideas here. My deep thanks to Paola and all of the students. I am also grateful to our guest speakers, especially Paul Horowitz, who remembered my article and spent two generous hours discussing SETI's modern prospects with our students.

In addition, I would like to thank the Experimental Study Group director Leigh Royden and associate director Graham Ramsay for buying into the seminar idea in the first place. Leigh also secured my appointment as a research affiliate in MIT's Department of Earth, Atmospheric, and Planetary Sciences, which came with a crucial benefit: library access.

Thanks as well to Mark Pelofsky for reading the manuscript and to my friends and colleagues at the Hub & Spoke audio collective for their support and encouragement.

In October 2018, Paola and I took our MIT seminar students to Harvard's Radcliffe Institute to attend "The Undiscovered," a day-long seminar organized by my friend the Harvard astronomer Alyssa Goodman. The event concluded with a talk by the astronomer Jill Tarter, the retired cofounder of the SETI Institute in California. Dr. Tarter is a celebrated SETI pioneer and was the model for the Ellie Arroway character in Carl Sagan's novel *Contact* (1985).[5] (Jodie Foster played the part in the film adaptation in 1997, the year after Sagan died.) Dr. Tarter's talk held us all rapt, and in a remarkable replay of my experience in October 1985, I was able to speak with her afterward and to tell her about this book. She was as gracious and engaging as I could have hoped.

So this volume is dedicated to Jill Tarter, Paul Horowitz, Carl Sagan, and all of the kind, lonely, and visionary scientists who have taught us how to look for extraterrestrials—and why the search itself connects us to the cosmos.

Cambridge, Massachusetts
Summer 2019

INTRODUCTION

Let's begin with a little counterfactual story. Imagine that the year is 1491. The place is somewhere south of Lake Ontario, in what is now upstate New York. Dozens of sachems, or chiefs, from a great league of Native American tribes, the Haudenosaunee, have convened a Grand Council meeting, and there is a startling item on their agenda.

They have learned from their shamans that a ceremonial pole of standard length casts a midday shadow that differs in length depending on whether the measurement comes from a village at the far northern edge of the league's territory or from a village at the far southern edge.

In each location, the pole points straight up—perpendicular to the ground. Such a difference could arise, the shamans reason, only if "up" is slightly different at each location on a north–south line, meaning that the poles are, in fact, oriented at slightly different angles to the sun's rays. The unavoidable implication is that the world is curved, forming a very large sphere. (The Greek mathematician Eratosthenes used the same reasoning to figure out Earth's circumference in the third century BCE.[1])

So large is this sphere that it must extend far beyond the lands known to the Five Nations and their neighbors. In fact, some of the shamans have observed, it's so big that there might be room for entirely new lands elsewhere on the sphere, occupied by unknown tribes—tribes who might want to trade with the Haudenosaunee or to fight them.

The conclusion seems incredible to the gathered sachems. But just in case, some of them argue, it might be worth preparing for the arrival of these supposed people from other lands.

No, others say. If other lands truly exist, their inhabitants would already have come across the oceans in boats to offer tribute or to make war. They have not. So surely the Haudenosaunee and the peripheral tribes are the only people in the world, and there is no cause for alarm.

Did I mention that the year is 1491?[2]

Where Is Everybody?

And now a second story, one retold so often among SETI scientists that it has entered into the realm of legend. It's the summer of 1950. In Los Alamos, New Mexico—site of the development of the first atomic bomb—a clutch of prominent nuclear physicists, including Hans Bethe, Edward Teller, Emil Konopinski, and Enrico Fermi, has reassembled to work on an even more powerful weapon, a hydrogen bomb.

The scientists gather daily for lunch at the Fuller Lodge, the main building of the old boys' school purchased by the US Army in 1943 to make way for the bomb laboratory. One day Herbert York, a visitor from the Physics Department at Berkeley, joins a table inside the lodge where Konopinski, Teller, and Fermi are in the middle of a conversation.[3]

Fermi possesses a brilliant and playful mind, and he enjoys posing semirhetorical questions that can be answered through rough estimation and back-of-the-envelope calculations. It's such a habit with him that his brain teasers—such as "How many piano tuners are there in Chicago?"—would come to be known as Fermi Problems or Fermi Questions. (The answer to the Chicago puzzler is a few hundred.)

On this particular day, the topic is not pianos, but flying saucers. Starting in 1947, there had been a series of highly publicized sightings of these unidentified objects. Press coverage had generated enough lighthearted buzz to inspire a cartoon in the *New Yorker*, showing aliens returning to their home planet toting trash cans stolen from the New York Department of Sanitation. (The cartoon was a tongue-in-cheek solution for two mysteries at once: UFOs and a rash of missing trash cans in New York.) Earlier in the day Konopinski had noticed this cartoon and mentioned it to Fermi, sparking the conversation.

Just as York sits down, Fermi bursts out: "Don't you ever wonder where everybody is?" Everyone at the table understands that Fermi is talking about extraterrestrials.

It's easy enough for the scientists to laugh off the popular idea that flying saucers are spaceships from other star systems, piloted by actual aliens. But that begs a larger question: If flying saucers aren't real and nobody has traveled through interstellar space to visit us, why not?

Fermi does some quick math, estimating quantities such as the age and size of the Milky Way, the number of stars and planets it contains, and the odds of intelligent life evolving on each planet. His calculation shows that our galaxy ought to be rife with civilizations. If no one has visited us, Fermi concludes, it's probably not because they don't exist; it's more likely that Einstein's universal speed limit—the speed of light, 3×10^8 meters/second—makes interstellar travel difficult or impossible.

The lunch companions nod and agree that the question is a deep and important one. But on this day they fail to resolve it. After lunch, they go back to designing their bomb.

The Power of Paradox

The point of these two stories—one a fable, the other true—is that we don't know what we don't know.

For thousands of years, the Native people of the Americas were safe in their isolation. It would have been natural to dismiss worries about rapacious, slave-driving, disease-carrying invaders—right up to the moment Columbus arrived.

Even today, there's no way to disprove Fermi's conclusion that we on Earth are safe in our isolation. He may have been right that the vast distances between the stars will keep us from ever meeting extraterrestrials. But Fermi was thinking only about *physical* contact. His conclusion would become moot if we were to detect an electromagnetic signal of intelligent extraterrestrial origin or notice some obvious sign of engineering elsewhere in the galaxy.

Over the years, Fermi's question "Where is everybody?" has ripened into a larger intellectual conundrum that students of SETI call the Fermi Paradox. The problem isn't really about the speed of light or whether the hypothetical aliens are too lazy or short-lived to visit us. It can be framed this way: everything we know about how planets form and how life arises suggests that human civilization on Earth should not be unique. In fact, our galaxy is old enough to have been thoroughly colonized, perhaps several times over. So we ought to see abundant evidence of extraterrestrial activity. But we don't see anything, not even radio blips and certainly not derelict spaceships or the monuments of dead civilizations. We should not be alone—yet apparently we are. How is that possible?

This question has fueled decades of debate, speculation, and, lately, some actual science. "It is hard to conceive a scientific problem more pregnant and richer in meaning and connection with the other 'Grand Questions' of science throughout the ages," writes Milan Ćirković in *The Great Silence*, a masterful new book showing how deep the problem really goes.[4]

Part of what makes the problem so deep is that it is in fact a formal paradox. The word *paradox* comes from the Greek term *paradoxon*, "contradictory opinion." It usually means "a proposition that proceeds from seemingly sound premises to a senseless or illogical conclusion."

Paradoxes can be seen as useful irritants. They beg for resolution. The key phrase in the definition is "*seemingly* sound premises." When a paradox pops up, it's often an indication that the premises are *not* as sound as

We should not be alone—
yet apparently we are.
How is that possible?

supposed or that there's something wrong in the chain of reasoning.

So what's the truth hiding behind the Fermi Paradox? Which flawed assumption would have to be excised to resolve it? What are we missing?

One possibility is that life, especially intelligent life, is less common than Fermi originally calculated. That's the idea favored by Peter Ward and Donald Brownlee, authors of *Rare Earth: Why Complex Life Is Uncommon in the Universe*.[5] We may find microbial life on many planets, Ward and Brownlee concede. But advanced, multicellular life can emerge, they propose, only when a planet enjoys an unusual combination of advantages, such as a Jupiter-size neighbor (to clear the neighborhood of debris), a large moon, plate tectonics, and a magnetic field.

Another possibility is that intelligent extraterrestrials do exist, but we just haven't met them yet. This is the answer offered by Carl Sagan in *Contact,* both the novel and the movie. In Sagan's fictional treatment, TV signals from Earth reach an alien detector near the star Vega, 25 light years away. This prompts an elaborate coded response that includes the blueprints for a kind of stargate; most of the suspense in the movie version is about who will get to go through it. The story amounts to an argument that we are not alone and that signs of intelligent life may lie just outside our grasp. We should keep looking for them because if we succeed, what we discover could change everything.

That's the solution to the Fermi Paradox that most SETI proponents still embrace today. But there are many, many other possible answers. In fact, one extremely useful book in this field is entitled *If the Universe is Teeming with Aliens ... Where Is Everybody? Seventy-Five Solutions to the Fermi Paradox and the Problem of Extraterrestrial Life.*[6]

The book you're holding is meant as a general introduction to the debate over the existence of intelligent extraterrestrials, and in the pages ahead we will encounter many ideas about how the Fermi Paradox may ultimately be resolved. In chapter 4, I boil the potential solutions down into general categories—fewer than 75, I promise—and review the arguments for and against each. Ward and Brownlee think that we *seem* to be alone because we *are* alone. Others, myself included, feel it's too early to come to that conclusion. At this point, we're still sorting out what former US Secretary of Defense Donald Rumsfeld, in a very different context, called the "known knowns," the "known unknowns," and the "unknown unknowns."

The entire field of astrobiology, for example, is dedicated to determining how life might arise in non-Earth-like environments. When the discipline was born several decades ago, scientists had little sense of the range of environments in which life might flourish, even here on Earth, let alone elsewhere in our solar system or elsewhere in the galaxy. But over the past 40 years, the field has advanced in fundamental ways. For one thing, we have discovered many types of "extremophiles," organisms thriving around undersea thermal vents and other places so inhospitable that no one would have expected to find life there, right up to the moment they did. On top of that, astronomers keep adding to the catalog of extrasolar planets, or exoplanets. As of this writing, there are 4,025 of them.[7] So far none are precisely Earth-like, but many do seem to orbit within the habitable zones of their star systems.

The point is that the more extremophiles and exoplanets we discover, the more room there is for exploration by astronomers and astrobiologists and the more unknown unknowns become known unknowns. With hard work, funding, and a little luck, we might even be able to convert some of them into known knowns. Watching this process fills me with wonder and hope.

Organizing Our Ignorance

Despite these discoveries, SETI remains an unusual corner of science, one where the blanks are even blanker than usual. The only hard piece of data we have is that the skies, so far, are silent. In 1975, the astronomer Michael Hart, in a famously skeptical article about SETI, called this silence "Fact A."[8] It's the observation that leads to the Fermi Paradox, and any serious argument for doing SETI research must grapple with it.

To restate the paradox: we don't see aliens, but we *should*, given a few seemingly reasonable assumptions. What are those assumptions? In the coming pages, we will meet the Drake Equation, first proposed by radio astronomer Frank Drake in 1960 as a way to estimate N, the number of technologically advanced and communicative civilizations there should be in the Milky Way. The equation helps to quantify the inputs that lead to the Fermi Paradox—in fact, Fermi used the same approach in his back-of-the-envelope calculations over lunch in Los Alamos. In the classic form of the Drake Equation, N is the product of seven factors, such as the number of stars in the galaxy, the fraction of stars that have planets, and the probability that simple life on a given planet will evolve into intelligent life. As we will see in the chapters ahead, we have learned a great deal about the first four factors in the equation, but we're still in the dark about the last three. The variable N could be exactly 1 (it certainly can't be less than that because we exist), or it could be much greater. We just don't know.

The Drake Equation isn't a traditional scientific equation, in the sense of a formula expressing how physical properties relate, like $E = mc^2$. But it was useful to Drake and his early SETI peers as a roadmap—a way to kick-start the discussion of the known unknowns. Indeed, the SETI pioneer Jill Tarter has called the equation "a wonderful way to organize our ignorance."[9]

This book proceeds in that same spirit, attempting to sort out the known knowns, the known unknowns, and the unknown unknowns. A conversation about aliens and the Fermi Paradox is by necessity partly conjectural. I won't hide my own opinions, and I'll try to make it clear when we're crossing the line from evidence-based reasoning to informed speculation—but I won't let that slow us down.

In a way, talking about SETI is like cooking stone soup. We're forced to start with little more than an idea. But if we borrow a bit of broth from the philosophers who have debated the plausibility of extraterrestrial life, some carrots from the oceanographers studying extremophiles, some seasoning from the astronomers seeking more exoplanets, and so forth, then we can probably make something intellectually nourishing.

Here's my list of ingredients. In chapter 1, I examine the surprisingly long history of speculation about aliens. When Aristotle said that nature abhors a vacuum, he was wrong; most of the universe is a near vacuum. But humans do seem to abhor the idea that we might be alone, and we have been debating the idea for thousands of years.

Scientists eventually realized that they could go beyond talk. Chapter 2 looks at the birth of SETI as a serious discipline in the 1960s. We'll learn how scientists turned to radio and optical techniques to begin the practical search for signals from extraterrestrial civilizations and how that search has evolved over the past 60 years.

Chapter 3 is about the revolutions in astrobiology and exoplanet research since 1977 as well as how the unexpectedly rapid progress in these areas has altered the way scientists think about the possibility of extraterrestrial life.

In chapter 4, armed with all this additional information, I return to the Fermi Paradox. Many intriguing solutions have been proposed, and I review them and assess their plausibility.

Finally, in chapter 5, I zero in on my favorite solutions and look at some new ideas for refocusing SETI work to increase the chances of resolving the paradox and finding extraterrestrials.

By the end, I hope you will agree that SETI is one of the most exciting and potentially world-changing research questions of our day and will feel inspired to keep exploring the subject on your own. You can do that with help from the sources cited in the notes, the glossary, and the list of further reading materials at the back of the book.

By the way, I'm going to assume that as of the moment you're reading this, aliens have *not* been discovered. If they have been, then the book is now a useless artifact of the precontact era. I would love nothing more, but I'm not that much of an optimist. So the paradox remains—and still begs to be unraveled.

ALIEN DREAMS

To feel small, all we have to do is look up. The sun, the Moon, the stars, the planets, and the Milky Way are evidence enough that Earth is not all that is. And for as long as humans have had words, we have been sharing stories about the presumed builders and occupiers of those vaulted heavens: the gods, spirits, angels, and demons who were, in a sense, the first extraterrestrials.

According to a Cherokee story, for example, the Milky Way is a great web spun across the sky by Grandmother Spider, who used it to reach the other side of the world and bring back the sun.[1] In one grisly Aztec myth, the war god Huitzilopochtli sprang from his mother Coatlicue's womb fully grown and fully armored. He beheaded his sister, Coyolxauhqui, who had been plotting to kill Coatlicue, and cast her head into the sky, creating the Moon.[2]

Materialist interpretations of the cosmos eventually began to take the place of mythological ones. But the idea that there might be other beings in the sky has stayed with us, and it found its first protoscientific roots in Greece in the sixth century BCE.

Anaximander, a philosopher who lived in Miletus in modern-day Turkey, contributed one key idea. He was the first to propose that Earth is a body floating in an infinite void, held up by nothing. For someone who lived 2,200 years before Isaac Newton, this was a stunning insight. The philosopher Karl Popper called it "one of the boldest, most revolutionary, and most portentous ideas in the whole history of human thought."[3] Anaximander also thought Earth was a cylinder with the continents arrayed on one flat end, so he wasn't right about everything. But he did invent the idea of space, a place with no absolute up or down. And just as important, Anaximander's system was the first to leave open the possibility that there are other worlds like ours. (Though, to be clear, he may not have believed that these worlds existed elsewhere in space. He may have thought they preceded or would succeed Earth in time or perhaps coexisted in some parallel universe.[4])

Anaximander's successors were more definite about the idea that came to be known as "the plurality of worlds" and more willing to explore its implications. In the fifth century BCE, the Thracian philosopher Leucippus and his

Anaximander was the first to propose that Earth is a body floating in an infinite void, held up by nothing.

pupil Democritus invented atomism: the belief that the visible universe consists of tiny, indivisible, indestructible atoms, churning in the void without purpose or cause. In this picture, worlds aren't divinely created; they simply form when enough atoms collide and stick together. Democritus thought that there was an infinite supply of atoms, so he reasoned that there must be an infinite number of worlds.[5] His pupil Metrodorus of Chios put it this way: "It seems absurd, that in a large field only one stalk should grow, and that in an infinite space only one world exist."[6]

And then there's Epicurus. He lived about a century after Democritus and is most notorious for his philosophy that pleasure—best obtained through modest, self-sufficient living—is the greatest good. But Epicurus read Democritus and thoroughly absorbed his empiricist, atomist worldview, including the idea that there must be many worlds. "There is an unlimited number of *cosmoi* [worlds], and some are similar to this one and some are dissimilar," Epicurus wrote in a letter to the historian Herodotus.[7]

Epicurus's ideas are important not just because they were prescient but because they became a long-lasting irritant for future philosophers and theologians. Unfortunately, most of his writings are lost. What we know about his thought comes mainly from *De rerum natura*, or *On the*

Nature of Things, a book-length poem by his Roman disciple Lucretius.

You can think of this book, written around 50 BCE, as the first volume of popular science. Here's what Lucretius said about the Epicurean view of other worlds:

> If store of seeds there is
> So great that not whole life-times of the living
> Can count the tale ...
> And if their force and nature abide the same,
> Able to throw the seeds of things together
> Into their places, even as here are thrown
> The seeds together in this world of ours,
> 'Tmust be confessed in other realms there are
> Still other worlds, still other breeds of men,
> And other generations of the wild.[8]

The passage is a milestone in discussions of extraterrestrials. It goes beyond the basic idea that an infinity must contain many worlds to offer what is probably the first straightforward assertion in Western literature that aliens must exist.

The first and sadly the last for a very long time.

The truth is that the mechanistic, nonsupernatural picture of the world offered by Anaximander, Democritus, and Epicurus was radical for its day. It failed to gain a large following in ancient Greece. In Athens in 450 BCE, the philosopher Anaxagoras posited that the sun is a fiery rock and that the Moon is an Earthlike body that glows in the sun's reflected light. He was promptly arrested on charges of impiety and sentenced to death. After his friend and former pupil Pericles came to his defense, he was released but banished.

Both Plato (428–348 BCE) and Aristotle (384–322 BCE) lambasted Democritus's idea of a plurality of worlds on theological grounds. Plato, a monotheist, argued that there is only one creator and that therefore there can be only one world, "if the created copy is to accord with the original."[9] Aristotle similarly thought that a plurality of worlds would require a plurality of Prime Movers to keep them in motion—a plainly heretical idea. The idea of infinite worlds also conflicted with his view of physics, in which the five basic elements—earth, air, fire, water, and divine aether—tend to move up or down toward their "natural places" at the center or the edges of the universe. Because things made of earth always sink to the center, Aristotle believed, Earth must be the only world, and there can be no solid bodies in the heavens.[10]

Though Aristotle was a pagan, his anthropocentric picture of the universe was a gift to early Christian theologians. The Book of Genesis, which says God purposefully created *the* heavens and *the* earth, left no room for other worlds or other sentient beings (unless you count angels and demons). The New Testament then introduced the idea that God was incarnated as Christ to rescue the faithful from sin and damnation—a flattering story implying that humans are uniquely worthy of Christ's sacrifice. As the scientist and Christian apologist William Whewell would later put it, the Incarnation made Earth into "the Stage of the Great Drama of God's Mercy and Man's Salvation."[11]

By contrast, Democritus, Epicurus, and Lucretius offered a picture of a purely mechanical universe where everything arises from the purposeless collisions of atoms and where humans might be just one of an infinite number of intelligent races. "Small wonder the early Christians tossed the Epicurean package, extraterrestrials and all, into the abyss of doctrinal errors," writes the Catholic ethicist Benjamin Wiker.[12]

As Christianity swept across the decaying Roman Empire in the third and fourth centuries CE, the Church Fathers ridiculed and suppressed the Epicureans and their ideas and allowed their writings to burn or crumble. Atomism, the pursuit of pleasure, the plurality-of-worlds idea—all of it slipped into darkness, where, as Wiker observes, "it stayed for nearly a thousand years."[13]

Nudged Aside

Somehow, though, Lucretius's poem *On the Nature of Things* managed to cross the abyss into the fifteenth century. *The Swerve*, a delightful book about "how the world became modern" by the Harvard literary scholar Stephen Greenblatt, tells the story of the Florentine book collector Poggio Bracciolini, who recovered a copy of the poem in the library of a monastery in southern Germany in 1417. Within 60 years, hundreds of manuscripts and print editions were in circulation, reigniting interest in Epicureanism.[14] Greenblatt argues that the poem's atheistic and materialist ideas helped usher in Renaissance humanism—an inquisitive philosophy that, despite its name, began to question humanity's privileged station in the cosmos.

Whether the credit is due to Bracciolini or not, the Renaissance saw steadily growing interest in the idea of the plurality of worlds and its corollary, the possibility that other worlds might be populated by other beings. Mikołaj Kopernik, better known as Nicolaus Copernicus, provided one key stepping stone.

The Polish mathematician and astronomer was born in 1473—coincidentally, the same year the first print edition of *On the Nature of Things* appeared. (Note the date here: Copernicus lived at the same time as Christopher Columbus, who was 22 years his senior; Leonardo da Vinci, who was 21 years older; Niccolò Machiavelli, four years older; and Martin Luther, 10 years younger.) Copernicus is central to the story of extraterrestrials not because he believed in them—the question didn't seem to interest him—but because he was the first person to propose, based on observation and calculation, that Earth was not the center of the visible universe.

Around 1510, Copernicus began writing the commentaries and manuscripts that would become *De revolutionibus orbium coelestium* (*On the Revolution of the Celestial Spheres*). Finally published in 1543, the year Copernicus died, the book upended the old Aristotelian system. It argued that Earth rotates around its pole, that the Moon orbits Earth, and that Mercury, Venus, the Earth–Moon system, Mars, Jupiter, and Saturn all travel around the sun at their own rates. Finally, it asserted that the firmament— the outermost celestial sphere, containing the stars— must be incomprehensibly far away, at least compared to the distances between the sun and the planets.

Copernicus's heliocentric model accounted for important oddities that the old Aristotelian system couldn't adequately explain, such as the occasional "retrograde" or backward motion of the other planets against the background stars. But, of course, heliocentrism wasn't immediately accepted, not least because it amounted to a huge demotion for Earth. It left us with only a single heavenly attendant, the Moon, and it forced Copernicus's readers to

Of course, heliocentrism wasn't immediately accepted, not least because it amounted to a huge demotion for Earth.

reckon with the idea that we live on a planet that is just like any other. This premise—that there's nothing particularly special about Earth and that we aren't in a privileged, central position to observe the universe—would come to be known as the Copernican principle, and it's at the core of the modern-day case for doing SETI research, as discussed in later chapters.

Copernicus knew his theory would provoke religious objections, which may be why he declined to publish it during his lifetime. His follower Giordano Bruno was not so cautious. Bruno was a Sicilian subject who entered the Dominican order in Naples and then became a religious vagabond. He read Lucretius and Copernicus, took their ideas deeply to heart, and made some startling leaps of his own.

In three sets of dialogues published between 1584 and 1591—*La cena de le ceneri* (*The Ash Wednesday Supper*), *De l'infinito universo et modi* (*On the Infinite Universe and Worlds*), and *De immenso* (Of vastness)—Bruno argued that at least some of the stars are suns with their own planets and that some of these planets must have their own residents. On this and many other subjects, Bruno's daring views conflicted with long-standing doctrines of the Catholic Church: for starters, that the universe was created for humanity alone and that there can be no people on other worlds without another Christ to redeem them. Bruno was arrested in Venice in 1592 on charges of blasphemy and heresy and sent to Rome, where his trial lasted seven years. On February 17, 1600, he was hanged naked upside down and burned at the stake.

Bruno's persecution was widely followed by people living outside Rome, but it couldn't prevent the emergence of a new understanding of the heavens. In 1609, Johannes Kepler, the German mathematician and astronomer, published *Astronomia nova* (*New Astronomy*), which extended Copernicanism in crucial ways. Understandably, Kepler was elated to receive a copy of Galileo Galilei's *Siderius nuncius* (*Starry Messenger*) soon after it was published the following year. The book announced Galileo's discovery of mountains on the Moon and four satellites orbiting Jupiter: we call them Io, Europa, Ganymede, and Callisto. These Jovian moons formed what was, in essence, a miniature solar system obeying the same rules as the planets. This discovery provided spectacular evidence for Copernicanism and in Kepler's mind confirmed his own theories about planetary motion. But here's the interesting part for our purposes: even though Kepler (a Protestant) knew of Bruno's travails and the Catholic Church's attitude toward the plurality-of-worlds idea, he sent Galileo (a Catholic) a congratulatory letter that included speculation about extraterrestrials. Any planet important enough to have moons, Kepler supposed, must also have people. "These four little moons exist for Jupiter, not for us," he wrote. "Each planet in turn, together with its occupants, is served by its own satellites. From this line of reasoning we deduce with the highest degree of probability that Jupiter is inhabited."[15]

Galileo cannily declined to endorse that idea. "The view of those who would put inhabitants on Jupiter, Venus, Saturn and the Moon, meaning by 'inhabitants' animals like ours, and men in particular" was "false and damnable," he wrote in his pamphlet *Istoria e dimostrazioni intorno alle macchie solari (Letters on Sunspots)* in 1613.[16] But while Galileo may have sidestepped Bruno's error in this case, he eventually ran afoul of the church for different reasons. His volume *Dialogo sopra i due massimi sistemi del mondo (Dialogue Concerning the Two Chief World Systems)*, a rousing defense of Copernicus, angered Pope Urban VIII and his inquisitors. In 1633, the church sentenced Galileo to a house arrest that lasted until his death in 1642.

So Many Earths

It's not my ambition in this chapter to mention every single scientist or philosopher who grappled with the question of extraterrestrials before the era of organized SETI research.[17] I'm only trying to demonstrate that the idea that other worlds might be home to alien beings—the word *alien* comes from the Latin term *alius*, "other"—has been part of our thought for as long as we have been looking skyward.

From Democritus to Galileo, thinkers treated this idea with great seriousness. After all, believing in aliens could get you banished or burned at the stake. But in 1686 a Frenchman named Bernard le Bovier de Fontenelle became the first writer to exploit the subject's humorous possibilities. His book *Entretiens sur la pluralité des mondes* (*Conversations on the Plurality of Worlds*) was another early example of science popularization.

Fontenelle made a rigorous case for Copernicanism, but to keep things entertaining he also used whimsical proto-science-fiction notions about the cultures of the other planets. The people of Venus, Fontenelle mused, are "sunburnt, full of verve and fire, always amorous, loving verses, loving music, inventing celebrations, dances, and tournaments every day." The inhabitants of Saturn, by contrast, are "quite phlegmatic. ... These are people who don't know what it is to laugh, who always take a day to answer the slightest question asked them."[18]

These ideas didn't contradict the doctrine of Christ's unique incarnation on Earth, Fontenelle reassured his readers, because people on other planets would not be descended from Adam and wouldn't need saving. Unfortunately, that didn't stop the church from putting *Conversations* on its Index of Forbidden Books.

Christian Huygens, the Dutch astronomer who explained the rings of Saturn and discovered its moon Titan, took a more serious tack in *Cosmotheoros*, published posthumously in 1698 and translated into English as *Celestial Worlds Discover'd; or, Conjectures Concerning the Inhabitants, Plants, and Productions of the Worlds in the Planets*. He noted that Venus and Jupiter have atmospheres, one requirement for life. He expanded on Bruno's assertion that other stars must have their own planetary systems and reasoned that where there are planets, there must be people.

> Why may not every one of these Stars or Suns have as great a retinue as our Sun, of Planets, with their Moons, to wait upon them? ... They must have their Plants and Animals, nay their rational Creatures too, and those as great Admirers, and as diligent Observers of the Heavens, as ourselves. ... What a wonderful and magnificent Scheme we have here of the magnificent Vastness of the universe! So many Suns, so many Earths, and every one of them stock'd with so many Herbs, Trees, and Animals, and adorn'd with so many Seas and Mountains![19]

By Huygens's time, the plurality-of-worlds concept was beginning to seem ordinary. Eighteenth-century thinkers such as Edmond Halley, Gottfried Leibniz, Alexander

Pope, Immanuel Kant, William Herschel, Pierre Laplace, and Thomas Paine accepted it as part of a scientific-realist worldview. This view was, however, still incompatible with strict Christianity. That's what motivated a leading nineteenth-century scientist and one-time believer in other inhabited worlds, William Whewell, to abandon pluralism and publish one of the strongest catalogs of scientific arguments *against* the idea.

A brilliant polymath, Whewell was a professor of mineralogy at the University of Cambridge, then a professor of moral philosophy, and finally the head of Trinity College, where Sir Isaac Newton had studied and taught. In the 1830s, Whewell published essays that left room for the idea of extraterrestrials. But he later grew increasingly disturbed by the question of whether God had provided "a like scheme of salvation" for every other world. If both pluralism and the Incarnation could not be true, Whewell decided he would stick with the Incarnation. So he assembled a scientific and philosophical broadside against the idea of other worlds, which he published in 1853 under the title *Of the Plurality of Worlds: An Essay*.

Whewell pointed out that humans, according to the geological record then being unearthed, had been present on this planet for only an "atom of time." If Earth had been, in effect, uninhabited through most of its history, then it wouldn't be surprising if other distant planets were also empty. In any case, he pointed out, no planets around other stars had yet been observed, and many nebulae, star clusters, and multiple-star systems would be unsuitable places for them. Here in the local neighborhood, Whewell noted, the Moon has no atmosphere or water; Jupiter features crushing gravity and may lack a solid surface; Saturn, Uranus, and Neptune are probably too far from the sun and therefore too cold to support life; and Mercury and Venus are probably too close to the sun and therefore too hot. He wasn't sure about Mars, but he theorized that only Earth is in what he called "the Temperate Zone of the Solar System."[20]

In short, though Whewell's ultimate goal was to defend Christian theology, he was the first to marshal empirical evidence to point out the real weaknesses in the plurality-of-worlds idea. This challenge was, in a sense, long overdue. Copernicus was correct to revoke Earth's privileges as the pivot point of the universe, but that insight by itself says nothing about what *else* might exist in the universe. We know today that Democritus and Epicurus were on the right track when they theorized about atoms and other worlds, but they didn't have any hard data, and neither did Bruno, Kepler, Huygens, or Fontenelle. Whewell concluded: "The belief that other planets, as well as our own, are the seats of habitation of living things, has been entertained, in general, not in consequence of physical reasons, but *in spite of* physical reasons." [21]

Though Whewell's ultimate goal was to defend Christian theology, he was the first to marshal empirical evidence to point out the real weaknesses in the plurality-of-worlds idea.

Coming from the master of Trinity, this attack caused a ruckus in the scientific world. Defenders of pluralism were forced to go back to their laboratories and telescopes (which is evidence, if you're in an optimistic mood, that materialists and religious believers aren't engaged in a winner-take-all war, but rather in a healthy competition of ideas). Even today, the essential aim of astrobiologists and exoplanet hunters is to provide what Whewell called the missing "physical reasons."

The Canal Builders

One of the researchers who poured new energy into the pursuit of extraterrestrials in the late nineteenth century was Percival Lowell. An amateur astronomer, Lowell used his wealth and his connections as a member of an old Boston Brahmin family to establish his own observatory in Flagstaff, Arizona, in 1894.

The year before that, the distinguished Italian astronomer Giovanni Schiaparelli had published *La vita sul pianeta Marte* (Life on Mars), laying out his observations of "seas," "continents," and waterways on Mars. After reading Schiaparelli's book and another on Mars by the French astronomer Camille Flammarion, Lowell became convinced that the alleged waterways were artificial canals, and he built the observatory in order to observe, document, and publicize them.

One piece of lore endlessly repeated in books, magazine articles, and web posts about Mars says that Lowell's imagination was fired by one of history's most comic mistranslations. Schiaparelli, so the story goes, described the lines he saw on the surface of Mars using the word *canali*, "channels." English translators, however, rendered it as "canals." A channel isn't necessarily artificial; a canal is. The misleading word choice was what supposedly sent Lowell on his wild quest.

This is one of those stories journalists call "too good to fact-check." In reality, Schiaparelli had begun talking about *canali* as early as 1878, the year after a close Mars–Earth approach. He was well aware that his work had inspired others to speculate that the *canali* were artificial, and perhaps used for irrigation. He did nothing to tamp down this speculation. "Their singular appearance and the fact that they are designed with absolute geometrical precision, as if they were drawn with a ruler or a compass, has led some to see in these features the work of intelligent beings, inhabitants of the planet," Schiaparelli wrote in *La vita sul pianeti Marte*. "I will be careful not to combat this assumption, which includes nothing impossible."[22]

Regardless of who inspired his canal obsession, Lowell set out to confirm Schiaparelli's discovery, making nearly nightly observations of Mars starting in mid-1894. He duly discovered 184 canals, putting Schiaparelli's 79 to shame. Lowell published these findings in a popular volume, *Mars* (1895), followed by *Mars and Its Canals* (1906) and *Mars as the Abode of Life* (1908). Like Schiaparelli before him, Lowell was struck by the "uniformity," the "precision," and the "supernaturally regular" appearance of the alleged canals. He wrote in the first volume, "Too great regularity is in itself the most suspicious of circumstances that some finite intelligence has been at work."[23]

Such a great collection of works would need builders, of course, and Lowell would go on to deduce—based on Mars's lower gravity—that Martians must be far larger and stronger than humans. And older and wiser, too. "A mind of no mean order would seem to have presided over the system we see—a mind certainly of considerably more comprehensiveness than that which presides over the various departments of our own public works," he wrote. "Certainly what we see hints at the existence of beings who are in advance of, not behind us, in the journey of life."[24]

The public greeted Lowell's work rapturously, scientists more coolly. Alfred Russell Wallace, the codiscoverer with Charles Darwin of evolution by natural selection, was still alive when Lowell's books appeared. He eviscerated the idea of intelligent, canal-building Martians. Wallace pointed out, correctly, that there is little liquid water on Mars to transport in canals.[25] And he anticipated later critiques of SETI by highlighting the fantastic odds against the appearance of even one technological species in a given star system, let alone two on neighboring planets. Given the series of evolutionary accidents that opened the way for the emergence of primates, each accident dependent on the previous one, "the total chances against the evolution of man, or an equivalent moral and intellectual being ... will be represented by a hundred million of millions to one," Wallace wrote.[26]

Wallace was right that there are no men on Mars. But there *was* an intelligence at work in the story: Lowell's.[27] We know from decades of telescopic, orbital, and robotic exploration of the red planet that there are no canals or even features such as sand dunes or dust storms that could create the illusion of canals. What Lowell saw had to have been what astronomer Simon Newcomb would call, in 1907, unconscious "visual inferences"—projections of Lowell's desire to see what he already believed was there. I'm reminded of the snide acronym sometimes used by tech-support workers to describe questions from naive computer users: PIBKAC, Problem Is between Keyboard and Chair. In Lowell's case, the problem was between the telescope eyepiece and the drawing pad.

But even before Wallace published his critique in 1904, it was too late to defuse Lowell's idea. Martians had escaped into popular culture. H. G. Wells took Lowell's concept of an ancient, advanced race of Mars dwellers and added a layer of imperial malice in *The War of the Worlds*, which was published in serial form in 1897 and as a print novel in 1898. Edgar Rice Burroughs used Mars, a.k.a. "Barsoom," as the setting for a series of pulpy stories and novels published between 1912 and 1948. Orson Welles adapted H. G. Wells's story as a live radio drama broadcast on Halloween Eve in 1938, and its simulated news format scared at least a few listeners into believing invaders from Mars had really arrived. The hostile-Martian cliché spread so quickly that by 1948 it would be satirized in the form of every nerd's favorite Looney Tunes villain, Marvin the Martian, followed in 1950 by Ray Bradbury's ground-breaking short-story collection *The Martian Chronicles*, about the conflicts between telepathic Martians and settlers from Earth.

Today we know that Mars is cold and dry and that if there are real Martians, they're probably microbes, buried below the surface. But Mars has been extremely fertile as garden for our own evolving theories, fears, and longings about extraterrestrials. We don't know yet whether the sky is full of "still other worlds with other breeds of men," as Lucretius poetically put it. Yet there remains one stubborn and absorbing fact: on the very next planet, life is not out of the question—even if that life winds up being us.

MAKING SETI INTO SCIENCE

For millennia, debaters could only conjecture and philosophize about the existence of extraterrestrials. The problem wasn't just that there was no physical evidence either way, as Whewell was right to point out. It was that nobody even knew how to get any (though Schiaparelli and Lowell deserve partial credit for at least putting their eyes to their telescopes).

All that began to change in 1959, when the British scientific journal *Nature* published a modest, three-page paper by the Cornell University physicists Giuseppe Cocconi and Philip Morrison entitled "Searching for Interstellar Communications."

Assume that a long-lived, technically advanced society exists somewhere in our corner of the galaxy, Cocconi and Morrison began. Assume they know our sun is of the type that could also support planets with life. If they wanted to send a message our way, how would they do it? And would we have the technology to detect it?

To answer the first question, Cocconi and Morrison reasoned that the hypothetical extraterrestrials would use radio waves, which travel at the speed of light and are able to pass through obstacles such as gas clouds and Earth's atmosphere. To answer the second question, they did a bit of math. They were able to demonstrate that if the aliens cranked up the power of their radio signal to a level slightly beyond our capabilities but presumably within theirs, then Earth-bound radio telescopes like those already being built in the late 1950s would be sensitive enough to detect it.

The paper amounted to an almost literal wake-up call. The interstellar party line might already be buzzing with conversation, Cocconi and Morrison were saying. Now that we have a phone, we ought to pick it up and listen. They concluded:

> The reader may seek to consign these speculations
> wholly to the domain of science-fiction. We
> submit, rather, that the foregoing line of argument
> demonstrates that the presence of interstellar signals
> is entirely consistent with all we now know and that
> *if signals are present the means of detecting them is
> now at hand*. Few will deny the profound importance,
> practical and philosophical, which the detection

Cocconi and Morrison reasoned that the hypothetical extraterrestrials would use radio waves, which travel at the speed of light and are able to pass through obstacles such as gas clouds and Earth's atmosphere.

of interstellar communications would have. We therefore feel that a discriminating search for signals deserves a considerable effort. The probability of success is difficult to estimate; but if we never search, the chance of success is zero.[1]

It's useful to step back and look at the mid-twentieth-century developments that cleared the way for Cocconi and Morrison's insights. In the 1930s, Morrison studied physics at Berkeley under Robert Oppenheimer, and during World War II he joined the Manhattan Project. At the University of Chicago's Metallurgical Laboratory, he worked on nuclear-reactor design, and later at Los Alamos he helped design the "explosive lenses" needed to ignite the first atomic bomb. He even transported its plutonium core to the Trinity test site in the back of his Dodge sedan. After the war, Morrison took a job at Cornell, became active in the nuclear nonproliferation movement, and got interested in the possibility of gamma-ray astronomy.

That's how he crossed paths with Cocconi, an Italian physicist who also taught at Cornell and studied cosmic rays—fast-moving, high-energy particles of matter that carry even more energy than gamma rays. Morrison knew that gamma rays could penetrate the interstellar dust that blocks our view of much of the Milky Way, and Cocconi knew that physicists were learning how to build synchrotrons that emit gamma-ray beams. Together, they wondered whether these beams could carry messages between stars. Unfortunately, the question was moot because the technology to precisely gather, focus, and measure gamma rays didn't yet exist. So their conversation shifted to a more promising possibility: radio frequencies.

The field of radio astronomy owed its accidental birth to a 26-year-old Bell Telephone Laboratories engineer named Karl Jansky. In 1932, Jansky was using a directional shortwave antenna to study the pesky radio static that interfered with transatlantic telephone connections. In the process, he discovered a mysterious radio signal that passed in front of his antenna every 23 hours and 56 minutes. This happens to be the length of a "sidereal day," or the time it takes for Earth to rotate once relative to the stars. (A solar day is slightly longer than a sidereal day because at the same time Earth is rotating around its axis, it's moving along its orbit around the sun. It has to rotate slightly more than once with respect to the stars to get back to the same Earth–sun orientation.) The sidereal period of the signal meant that the source had to be in the sky, but it couldn't be the sun. Jansky eventually tracked the signal to the constellation Sagittarius in the densest part of the Milky Way and thereby became the first person to discover a radio-emitting object outside our solar system.

It's now thought that the signals Jansky found are emitted by electrons trapped in the magnetic field of Sagittarius A*, the supermassive black hole at the center of the galaxy. In honor of the discovery, the fundamental unit of irradiance in radio astronomy was later named the jansky. But Bell Telephone—which was, after all, in the communications business, not the astronomy business—didn't give Jansky time to follow up on his find. As a result, the new science of radio astronomy lay mostly fallow through the years up to and including World War II.

The war, however, utterly transformed the field, as it did so many others, from nuclear physics to computing and rocketry. In parallel with the Manhattan Project, the United States and Britain undertook a crash project to use microwave-radio pulses to help detect incoming attackers and to guide bombers to their targets. In the process of developing radar, scientists at MIT's Radiation Laboratory also made drastic advances in radio electronics, such as techniques for filtering out receiver noise. As soon as the war was over, astronomers realized that the new technologies would make it possible to detect even weak radio sources in the sky.

German technology helped, too. British scientists using Wurzburg dishes, antiaircraft radar left behind by the Nazis when they fled the coasts of France and Holland, were the first to observe radio emissions from sunspots. And in the Netherlands, astronomer Jan Oort used a Wurzburg dish to corroborate the discovery in 1951 of the 21-centimeter "hydrogen line."

Because the hydrogen line is key to the SETI story, I'll slow down to explain it. A hydrogen atom consists of one proton and one electron. In quantum mechanics, both particles have a type of angular momentum called *spin*. When the two particles have the same (parallel) spin, a hydrogen atom has slightly more overall energy; when they have the opposite (antiparallel) spin, the atom has less energy. Once in a great while—every 10 million years on average—the electron in a high-energy hydrogen atom can flip its spin from parallel to antiparallel. When that happens, the atom gives off a burst of radio energy at an extremely precise wavelength (21.1061 centimeters) and frequency (1420.4058 megahertz [MHz]): the hydrogen line.

Although flipping is extremely rare for an individual hydrogen atom, interstellar gas clouds contain so many hydrogen atoms that a few electrons are always flipping and some waves are always leaking out at roughly 21 centimeters. The Harvard physicists Harold Ewen and Edward Purcell were the first to detect this energy in the spring of 1951, followed immediately by Oort.

Because the hydrogen line is so narrow and precise, it's easy to find it against other background radiation, which means it's also easy to measure how the line shifts toward the red end of the electromagnetic spectrum when the source of the emission is moving away from the detector or toward the blue end if it's moving toward the detector. (The same Doppler shift makes a car's horn pitch up when it's coming at you and down when it's going away from you.) Oort used this effect to make the first radio maps of the spiral arms of the Milky Way. He also proved that the galaxy as a whole is rotating and that our solar system is nowhere near its center. In fact, we're on the inner rim of a minor spur of the Orion Arm, an unremarkable structure about halfway between the galaxy's center and its outer edge.

Astronomers in the United States, energized by these discoveries, convinced the National Science Foundation to finance a cluster of much larger radio telescopes. Construction of these telescopes began in 1958 at the National Radio Astronomy Observatory in Green Bank, West Virginia, in the Allegheny Mountains. The work was considered to be so important that the Federal Communications Commission created the National Radio Quiet Zone around Green Bank. To this day, broadcasters within this 13,000-square-mile area must operate at greatly reduced power, making observations easier; close to the observatory, even microwave ovens and Wi-Fi routers are prohibited.

Before Cocconi and Morrison collaborated, they knew about the boom in government-funded telescope building. Their *Nature* paper amounted to a proposal that the new instruments be used, at least some of the time, to search for signals from extraterrestrials. Thanks to what Ewen, Purcell, and Oort had found, they even had some ideas about which radio channels an alien race might use to make their messages easy to find.

> At what frequency shall we look? Just in the most favored radio region there lies a unique, objective standard of frequency, which must be known to every observer in the universe: the outstanding radio emission line at 1,420 Mc./s. (λ = 21 cm) of neutral hydrogen. It is reasonable to expect that sensitive receivers for this frequency will be made at an early stage in the development of radio-astronomy. That would be the expectation of the operators of the assumed source, and the present state of terrestrial instruments indeed justifies the expectation.[2]

Mathematician Style

Cocconi and Morrison weren't the only scientists thinking along these lines. In early 1959, a young Frank Drake got a job at Green Bank, where the first instrument, a 26-meter dish called the Tatel Telescope, had just been completed. Not much earlier, while finishing his graduate studies at Harvard, Drake had pointed a smaller radio telescope at the Pleiades star cluster and detected what seemed to be a narrowband signal of intelligent origin. He quickly determined that the signal was coming from Earth, not the Pleiades, but he couldn't forget the thrill of that moment—he had been bitten by the SETI bug. At Green Bank, he talked the new director into letting him use the Tatel dish to look for more signals. And he knew where he wanted to look: the 21-centimeter line.[3]

By the time Cocconi and Morrison's paper came out in *Nature* that September, Drake and his collaborators were already building the amplifiers and other equipment they would need for Project Ozma, named after the benevolent princess of Oz in L. Frank Baum's novels. From April to July 1960, Drake collected 150 hours of radio readings from the vicinity of Tau Ceti and Epsilon Eridani, sunlike stars located 12 and 10.5 light years away from Earth, respectively.

Aside from a few false positives—a plane passing overhead, a signal from a secret military radar facility—Project Ozma detected nothing unusual. (In retrospect, though, Tau Ceti was a smart choice of targets. It's now thought that this star has four or five planets, each with two to four times Earth's mass. Epsilon Eridani has two asteroid belts and possibly two planets.) The project did, however, succeed in attracting the attention of *Time* magazine as well as that of J. Peter Pearman, a staffer for the Space Science Board at the US National Academy of Sciences. In the summer of 1961, Pearman called Drake and said he wanted to organize a conference aimed at drumming up more federal support for the search for extraterrestrials.

Drake immediately agreed to help. But whom should they invite? "We put our heads together to name every scientist we knew who was even thinking about extraterrestrial life in 1961," Drake later wrote. "That turned out to be all of ten people, including Pearman and me." The other eight were Cocconi and Morrison; the electronics entrepreneur Dana Atchley; the Hewlett-Packard researcher Barney Oliver; the neuroscientist John Lilly; the chemist Melvin Calvin; the director of the National Radio Astronomy Observatory, Otto Struve; and Carl Sagan, who was only 26 at the time but, according to Drake, "knew more about biology than any astronomer I had met."[4]

With the addition of one more participant, NASA astrophysicist Su-Shu Huang, the group met at Green Bank in November. It fell to Drake to prepare the agenda. He asked himself: What would scientists need to know to assess the likelihood of success for future Project Ozmas? And he began listing topics, some of which could be boiled down to quantities or fractions.

> Surely we needed to know the number of new stars born each year. If there were planets out beyond the Solar System that were suitable for life, how many of them would actually become homes to living things?

How many of those things might be intelligent? ... Then it hit me: the topics were not only of equal importance, they were also utterly independent. Furthermore, multiplied together they constituted a formula for determining the number of advanced, communicative species. I quickly gave each topic a symbol, mathematician style, and found that I could reduce the whole agenda for the meeting to a single line.[5]

In this way, Drake worked out the equation that would forever bear his name. He wrote it on a chalkboard during the opening session, and the attendees spent the next three days trying to come up with believable estimates for each term.

The equation looked like this:

$$N = R^* f_p n_e f_l f_i f_c L$$

In Drake's formulation, R^* is the rate of formation of stars in the Milky Way that could support life, f_p is the fraction of stars that have planets, n_e is the number of planets per star that are friendly to life, f_l is the fraction of those planets where life emerges, f_i is the fraction of those planets where intelligent organisms evolve, f_c is the fraction of intelligent species that develop technology for interstellar communication, and L is the length of time such a species remains detectable—in effect, how long they last before dying or snuffing themselves out. (For math nerds: If R^* is expressed in stars per year, and L is expressed in years, then those units cancel out, and N is a unitless integer—the number of communicative civilizations in the galaxy.)

When the Green Bank group plugged in their best estimates for the terms of the equation, they were surprised to find that the product of the first six terms, R^* through f_c, came out to 1. "Thus the value of N seemed to hinge solely on the value of L," Drake later wrote. In other words, if you assumed that an advanced technological civilization would last 100,000 years, then the equation would say that there should be 100,000 communicating civilizations in the galaxy. Drake recalled that at the end of the meeting Struve offered a toast: "To the value of L. May it prove to be a very large number."[6]

The Green Bank conference attendees, by the way, seemed to be aware of the simultaneously historic and outlandish nature of their discussions. The question of dolphin smarts kept coming up—if intelligence had emerged on Earth not once but twice, it might skew the estimate for f_i. As a consequence, the attendees would later come to call themselves the "Order of the Dolphin."

In 1961, filling in the equation required a great deal of pure guesswork on the Dolphins' part. Only R^*, the rate of star formation in the Milky Way, was well understood at the time. The group pegged it at one star per year. That

If you assumed that an advanced technological civilization would last 100,000 years, then the equation would say that there should be 100,000 communicating civilizations in the galaxy.

was close enough: astronomers now think around 7.5 stars are formed per year.

Moving into the realm of pure conjecture, the Dolphins estimated that the value of f_p is 0.2 to 0.5, meaning one-fifth to one-half of sunlike stars would have planets. They thought n_e, the number of Earth-like planets per star, is between 1 and 5. In an act of extreme optimism, they assigned a value of 1 to both f_l and f_i, meaning that 100 percent of these Earth-like planets would develop life and that intelligent organisms would emerge on these planets 100 percent of the time. They put f_c at 0.1 to 0.2, meaning that 10 to 20 percent of intelligent species would eventually figure out how to build radio telescopes. And they guessed that such species would last between 1,000 years and 100 million years.

When they did the math, N, the number of communicating civilizations in the galaxy, came out to at least 20 and at most 50 million.

Now, it's worth noting that these results don't help to solve the Fermi Paradox. In fact, they make it more puzzling. Even if there are only 20 advanced civilizations in the galaxy, then at least one of them should have visited our solar system by now if you accept Fermi's assumption that societies always expand to fill the available space. At the other extreme, if there are 50 million civilizations, signs of their existence should be as visible as the litter along a highway.

But we'll come back to this point later. The real purpose of the Drake Equation was not to generate an ironclad estimate for N. It was simply to sketch the kinds of questions scientists would need to study in order to get a handle on the processes that give rise to technological civilizations. These questions clearly went beyond astrophysics to encompass planetary science, biochemistry, evolutionary biology, and cultural studies. In that sense, the equation set out a bold research agenda for the future of SETI science.

It also functioned as a public-relations tool. It gave astronomers an easy way to explain SETI to nonscientists. And if the Dolphins could make a credible case that N is large, it would help to justify further spending on SETI.

Soviet SETI

If Pearman and Drake's list of "every scientist we knew who was even thinking about extraterrestrial life" amounted to only 10 people, it was because they didn't know enough scientists. Three years after the Green Bank conference, scientists in the Soviet Union held their own much larger meeting, the First All-Union Conference on Extraterrestrial Civilizations and Interstellar Communication, at the Byurakan Astrophysical Observatory in Armenia. Its organizer was the Ukrainian astrophysicist and radio astronomer Iosif Shklovskii, who has been called "the Soviet Frank Drake." And one of its stars was Shklovskii's student Nikolai Kardashev.[7]

Although Shklovskii and Kardashev felt that the number of other civilizations in our galaxy might be small, they were more willing than the Green Bank group to speculate about the nature of those civilizations. "We are only infants as far as science and technology are concerned," Shklovskii said in his opening talk. In tune with Marxist philosophy about the inevitability of social evolution, he believed that other civilizations would have technology far beyond ours. How far? At Byurakan, Kardashev proposed a way of classifying civilizations according to the scale of the energy resources they put to use. Type I civilizations can use all of the energy available on their planet. Humans are at or near Type I. Type II civilizations use all of the energy available from their home star. (A "Dyson sphere," first envisioned by physicist Freeman Dyson in 1960, is a hypothetical megastructure built around a star to capture all of its energy; it is thus a perfect expression of Type II technology.) A Type III civilization can control the energy of an entire galaxy.

If civilizations follow this path of evolution, it ought to make the job of SETI scientists easier, Kardashev reasoned, because the activities of Type II and Type III civilizations should be detectable from Earth. The fact that no such activity is visible might point to a low value for *L*—meaning that most civilizations destroy themselves somewhere between the Type I and Type II stages. In the early 1960s, this possibility seemed all too real. Nonetheless, after the Byurakan meeting a group in Moscow led by Kardashev and another group in Gorky (now Nizhny Novgorod) led by V. S. Troitsky began a series of all-sky searches for extraterrestrial signals. They used nondirectional antennas and sampled a broad range of frequencies on the assumption that Type II or Type III supercivilizations would be sending us short, obvious "call signs" at extremely high power.[8]

Project Cyclops

Although Pearman and the other Dolphins had hoped the Green Bank conference would accelerate SETI work in the United States, in reality the discipline cooled to a low simmer. Space scientists focused on more pressing matters, such as beating the Soviets to the Moon and sending the first robotic probes to other planets.

Sagan got busy helping NASA with its Mariner 2 mission to Venus. Even so, he found time for a collaboration with Shklovskii. Sagan expanded the Ukrainian's book *Vsellennaia, Zhizn, Razum* (*Universe, Life, Mind,* 1962), which had provided a foundation for the Byurakan meeting, and published it in the United States in 1966 under the title *Intelligent Life in the Universe*. It sold well and became the first public SETI manifesto (unless you count Fontenelle's work *Conversations on the Plurality of Worlds* in 1686—also a best seller in its day).

The book strained to portray the search for extraterrestrials as a credible branch of science, despite the amount of conjecture still involved. "Is it in fact possible to call a book dealing with intelligent life in the universe 'scientific'?" Shklovskii and Sagan asked. "We are deeply convinced that the problem can be approached responsibly only if the assumptions involved are stated explicitly, and if the most efficient use of the scientific method is made. Even then, we shall not come to many final answers. But the formulation of the problems has itself significance and excitement."[9]

Five years later NASA asked Barney Oliver—one of the original Dolphins—to figure out how much time, hardware, and money would be required to "mount a realistic effort … aimed at detecting the existence of extraterrestrial (extrasolar system) intelligent life."[10] Oliver interpreted his mandate broadly, ultimately producing a 250-page argument that extraterrestrials were probably waiting to be found if we could only start looking in earnest. There is nothing special about Earth or its people, the report argued: "The basic processes of stellar, chemical, biological, and cultural evolution are universal and, when carried to fruition, lead to technologies that must have close similarities to ours today and in the future."[11]

Following the same reasoning as Cocconi and Morrison, Oliver's report concluded that the microwave-radio band would be the best place to look for an extraterrestrial beacon. To have a decent chance of succeeding, given the huge number of locations and frequencies that would need to be examined, the report recommended that NASA build a collection of steerable radio-receiver dishes. Oliver's plan mimicked New Mexico's Very Large Array—which was already in the planning stages and would ultimately consist of 27 dishes, each 25 meters in diameter, placed along three tracks, each 21 kilometers long—in that it would use a process called interferometry to simulate an even larger receiver. The big difference was that Oliver's facility, dubbed Cyclops, would have 1,000 to 2,500 dishes, each 100 meters in diameter. The installation would be so massive that it would require a new city (named "Cyclopolis," of course) to service its staff and their families. The price tag: $6 billion to $10 billion, not even counting the need for a second Cyclops on the opposite side of the planet for full sky coverage and continuous monitoring.

From one point of view, the extravagant Project Cyclops proposal was a product of its time. NASA had just met President John F. Kennedy's insanely ambitious schedule for a Moon landing. At the time, it did not seem improbable that space hotels, Moon colonies, interplanetary spaceships, and sentient computers like those depicted by Stanley Kubrick in the film *2001: A Space Odyssey* (1968) would actually be built within a few decades. NASA had asked Oliver to think big, and he had.

But with the Apollo missions nearly over, funding for NASA was already declining fast. Project Cyclops would have consumed about one-fifth of the agency's 1971-level budget for the next decade. NASA distributed 10,000 copies of Oliver's report, but in reality the project was an instant no-go. "Although NASA intended the Cyclops Report to make SETI seem doable, the team actually accomplished the opposite," remarks journalist Sarah Scoles.[12]

Speaking for Earth

Instead of listening *for* extraterrestrials, SETI scientists spent much of the 1970s engaged in a quixotic, largely symbolic, and far less expensive effort to talk *to* extraterrestrials. This pursuit is sometimes called "communication with extraterrestrial intelligence," CETI, or "messaging extraterrestrial Intelligence," METI. It was pushed along by a joint US-Soviet meeting at the Byurakan Astrophysical

Observatory in 1971, sometimes called Byurakan II, which brought together all the field's luminaries—Morrison, Drake, Sagan, Dyson, Shklovskii, Kardashev, and even the artificial-intelligence pioneer Marvin Minsky and the DNA codiscoverer Francis Crick—to ask what common language might be developed to communicate with advanced civilizations. Quite naturally for a group of scientists, they concluded that the universal language was arithmetic. "It is probably easier to communicate with a Jovian scientist than with an American teenager," Minsky joked.[13]

Sagan soon had an opportunity to put the idea into practice. In 1972 and 1973, NASA launched the Pioneer 10 and Pioneer 11 probes on paths that would take them past Jupiter and Saturn. Sagan was aware that the probes would become the first two human-made objects to leave the solar system, so he asked NASA for permission to attach a small plaque to each craft, designed for the benefit of a civilization that might at some point in the distant future intercept one of them.

The agency agreed. The gold-anodized 23-by-15-centimeter plaques, designed by Sagan and Drake and illustrated by Sagan's then wife Linda Salzman Sagan, include an illustration of the hydrogen spin-flip transition—meant to define time and distance scales—as well as a diagram of the solar system's location relative to 14 pulsars, identified by their frequencies. In effect, it was a math geek's map back to Earth. The plaques also show two nude human figures, a man and a woman. They're unlikely ever to be viewed by alien eyes, but the figures have generated endless controversy here on Earth. Some commentators saw them as pornographic, while others have perceived racial and gender bias at work in the Sagans' drawing in that the figures can be seen as European, and only the man's hand is raised in greeting.

In 1974, Drake upped the ante. By that time, he was director of the National Astronomy and Ionosphere Center, which ran the 10-year-old Arecibo radio telescope in Puerto Rico. At a ceremony marking the completion of a major project to make the telescope more sensitive and more hurricane proof, Drake used the 305-meter dish to transmit a short radio message encoding even more information about Earth and its inhabitants.

The Arecibo message consisted of 1,679 binary digits: the product of two prime numbers, 23 and 73. If the receiver of the message were to arrange the "on" and "off" bits in a grid 23 squares across and 73 squares deep, a crude diagram would emerge in a style reminiscent of the simple two-dimensional graphics of video games of that decade, such as Pong. The diagram shows our decimal number system, the atomic numbers of the elements in human DNA (hydrogen, carbon, nitrogen, oxygen, and phosphorus), the chemical formulas for the sugars and bases in DNA nucleotides, a graphic of the DNA double helix, a blocky picture of a human (it looks more like a killer robot), the height of an average male, the human population of Earth (4.3 billion at that time), a graphic of our solar system, and a picture of the Arecibo telescope.

The message was beamed out at extremely high power—enough to be detectable by an Arecibo-size telescope anywhere in the galaxy. But it's unlikely to be received by anyone, given that it was aimed narrowly at M13, a globular star cluster 25,000 light years away; by the time the message gets that far, the cluster will have moved out of the way. According to the Cornell astronomer Donald Campbell, who was a research associate at Arecibo at the time, the Arecibo message "was strictly a symbolic event, to show that we could do it."[14] Drake says the message was merely meant to be "eye-catching" and "spectacular."[15]

But at the same time, it was a kind of fantasy template for an *incoming* message. If we ever pick up an intelligent radio signal from the stars, it would be extremely convenient if it were encoded using a mathematical scheme that SETI researchers could recognize and decipher, such as the prime-number grid. But perhaps predictably the Arecibo message, too, sparked controversy. Days after the ceremony, Martin Ryle, the royal astronomer of England, wrote to Drake to complain that it was "very hazardous

"This is a present from a small, distant world, a token of our sounds, our science, our images, our music, our thoughts and our feelings."

to reveal our existence and location to the Galaxy; for all we know any creatures out there might be malevolent—or hungry."[16]

The prospect of alien invasion didn't deter the SETI scientists. They reasoned that we had been leaking radio and television broadcast signals to the rest of the Orion Arm since the 1930s, so what difference could a short, deliberate broadcast make?

Indeed, when the next big opportunity to send an outgoing message came along, Sagan took it. He led a NASA committee that chose 115 still pictures and an hour of sounds, voices, and musical passages for the Voyager Interstellar Record, which was attached to the Voyager 1 and Voyager 2 probes before they were launched in 1977. (Both probes have now left the solar system and will speed endlessly into the deep space between stars.) The record was meant as a treasure trove for alien anthropologists in the distant future. But its contents, copies of which have been shared widely here on Earth since 1977, also functioned as a kind of time capsule. "This is a present from a small, distant world, a token of our sounds, our science, our images, our music, our thoughts and our feelings," President Jimmy Carter said in a statement included in the record. "We are attempting to survive our time so we may live into yours."

Wow!

After Project Cyclops proved dead on arrival at NASA, scientists determined to keep looking for extraterrestrials had to look outside the agency and devise more affordable approaches. One obvious idea was to repurpose, or piggyback on, existing radio telescopes.

Even before Cocconi and Morrison's paper, the Ohio State University radio astronomer John Kraus won funding to build a wacky-looking radio telescope in Delaware, Ohio, called Big Ear. With a large reflecting vane at one end of a large, open field and a curved focusing reflector at the other end, it acted like a giant flatbed scanner. Earth's rotation dragged the telescope's field of view across the sky, with no point remaining in view for more than 72 seconds. Big Ear was used from 1965 to 1972 to compile the Ohio Sky Survey, a catalog of 20,000 radio sources in space, most of which had never been seen before. After Congress defunded the survey, Kraus decided to start looking for signals of intelligent origin. Recording began in December 1973, concentrating on the 1,420 MHz channel—the hydrogen line.

Almost four years later, on August 15, 1977, Big Ear recorded a strong signal coming from a point near the globular cluster M55 in the constellation Sagittarius. The signal was at the hydrogen frequency; it was extremely strong (30 times louder than the background noise); and it

Big Ear was used from 1965 to 1972 to compile the Ohio Sky Survey, a catalog of 20,000 radio sources in space, most of which had never been seen before.

wasn't local (it gained intensity for 36 seconds and waned for 36 seconds, just as would be expected for an astronomical object crossing Big Ear's field of perception). When the astronomer Jerry Ehman reviewed a printout of the observing session three days later, he circled the signal's peak with a red pen and wrote "Wow!"

Forevermore known as the "Wow! signal," Big Ear's recording was one of SETI's most tantalizing moments.[17] But the next time Ehman scanned the area, the signal wasn't there. Multiple studies of the M55 region in the years since then have had the same null result. It's still possible that the Wow! signal was of intelligent origin— astronomers haven't come up with a persuasive natural explanation. But in that case it's hard to see why the senders wouldn't have repeated the message, and a message that isn't replicated has little credibility. (In a way, this problem exposes a weakness in our whole approach to SETI: we can't yet listen to the whole sky all the time, so we depend on putative extraterrestrials to help us out by building long-lived beacons.)

In the years that followed, the Ohio State team picked up nothing remotely as interesting as the Wow! signal. In 1998, just as Big Ear was entering the Guinness Book of World Records as the home of history's longest continuous SETI search, the telescope was demolished to make way for a private golf course.

Small and Stubborn

When the Berkeley astronomer Stuart Bowyer read the Cyclops report, it gave him an idea similar to Kraus's. Why not build a SETI system that would be symbiotic with existing radio-astronomy efforts? If Bowyer could get a copy of incoming radio data to be analyzed later for suspicious signals, other astronomers could carry on with their science. He called the idea "piggybacking," and he found an astronomer at Hat Creek Radio Observatory in northern California, Jack Welch, who was willing to let him try it out. Jill Tarter, an astronomy graduate student at Berkeley at the time, volunteered to help (she and Welch would later marry).

They called their project the Search for Extraterrestrial Radio Emissions from Nearby Developed Intelligent Populations, or SERENDIP.[18] It was one of the first to listen to the sky on multiple frequencies simultaneously: just 100 channels at first, covering a band of frequencies 100 kilohertz wide, but many more later. SERENDIP was switched on in 1979 and has gone through multiple upgrades and migrations since then, from Hat Creek to Green Bank and ultimately to Arecibo, where it's still in operation today.

SERENDIP got off the ground with little government funding, was relatively small in scale, and stayed aloft thanks mostly to the stubbornness and entrepreneurship of researchers who believed in the importance of the search. In other words, it was typical of organized SETI efforts from the late 1970s on. At that point, the expansive, adventurous spirit of government-funded space science was dormant. Radio astronomy was no longer a novelty, and SETI's essentially speculative nature—and its proximity to UFOs and Martians in the popular imagination—made it vulnerable to political attacks. SETI researchers had to fight for every penny of federal funding or go their own way.

Jill Tarter did both. In 1975, NASA started funding a small design study, the Microwave Observing Project (MOP), to determine what kind of technology would be needed to keep SETI going—in essence, Cyclops on the cheap. Tarter, who was by then on the faculty at Berkeley, contributed hardware designs. But MOP had enemies in Washington, DC, notably William Proxmire, a Democratic senator from Wisconsin with a reputation as a budget hawk. He gave the project his Golden Fleece Award in 1978, arguing that "there is now not a scintilla of evidence that life beyond our own solar system exists." (Discovering such evidence was, of course, the whole point.) When the bad publicity didn't put an end to the project, Proxmire tried to slice it from NASA's budget in 1981, only to be personally talked out of doing so by Carl Sagan.

After that, MOP limped along on a minimal budget. In 1984, Tarter came up with a plan for reducing overhead costs and making NASA's money go further by founding the nonprofit SETI Institute in Mountain View, California, near the NASA Ames Research Center. The institute took over development of the basic hardware needed to accelerate radio SETI—especially spectrum analyzers that could process signals on many frequencies at once.

A plan slowly emerged to put these analyzers to work in two separate searches: a deep search targeting stars within 100 light years and a shallower survey of the whole sky. In 1992—by which time MOP had been rechristened the High Resolution Microwave Survey (HRMS)—the equipment for the deep search was finally ready to be airlifted to Arecibo. The survey began there on Columbus Day, the five hundredth anniversary of the "discovery" of North America.

But HRMS didn't last long. Within a year, Proxmire's successors in Congress managed to zero out its $12.3 million budget, killing SETI work at NASA once and for all. "As of today, millions have been spent and we have yet to bag a single little green fellow," read a mocking press release from the office of the main ax wielder, Nevada senator Richard Bryan. "Not a single Martian has said 'take me to your leader,' and not a single flying saucer has applied for FAA approval." Tarter would have to look elsewhere for funding.

Back East, meanwhile, the physicist Paul Horowitz had chosen a SETI strategy that largely eschewed federal funding and all the hazards that come with it. An electronics whiz, Horowitz devised a spectrometer that could measure signal strength on 65,000 extremely narrow radio channels around the hydrogen line. (The narrower the width of each frequency examined, the more easily the detectors could discriminate signal from noise and the less interference there would be from Earth-bound sources.) In 1978, Horowitz used the spectrometer at Arecibo to examine 185 nearby stars.[19] With support from Sagan's science-advocacy group the Planetary Society, he then improved and miniaturized his equipment into a portable device called Suitcase SETI, which was tested at Arecibo and adapted again for installation at the Harvard–Smithsonian radio telescope in the rural town of Harvard, Massachusetts.

Between 1983 and 1985, in an operation called Project Sentinel Horowitz's team completed several scans of the northern sky. But the Sentinel spectrometer had a couple of key limitations. It couldn't account for potential Doppler shifts in the incoming signal, meaning that if Horowitz hoped to pick up signals around the magic 1,420-MHz frequency, the hypothetical transmitting civilization would have to correct their signal *in advance* for the relative movement of their star system and ours. Also, the Sentinel processors were slow: it took at least 30 seconds to analyze incoming data for a given spot in the sky on all 65,000 frequencies, by which time the telescope would have moved on to the next target.

To get past these problems, Horowitz designed a next-generation spectrometer with more computing power. The new Project META (Megachannel Extraterrestrial Assay) machine could scan 8.4 million channels near the hydrogen line in real time. It could also correct for Earth's rotation and for other motions such as the solar system's movement around the galactic center—meaning that for the first time researchers might be able to detect a narrowband signal that had not been tailored specifically for Earthlings.

Horowitz built the Project META spectrometer with help from the Planetary Society and a $100,000 gift from the filmmaker Steven Spielberg. Sagan and Spielberg visited the Harvard campus in September 1985 for a symposium celebrating Project META's launch. And that's where I come into the story briefly. In my first outing as a student journalist, I met Sagan and interviewed Horowitz (see the preface). I couldn't help asking Horowitz the standard question every journalist asks of every SETI scientist: What are the chances you'll succeed? "Nobody knows," he answered. "We can use equations to guess at the number of extraterrestrial civilizations that should exist in our galaxy. There's no way to be perfectly accurate, but the numbers show they should be out there."[20]

"We can use equations to guess at the number of extraterrestrial civilizations that should exist in our galaxy. There's no way to be perfectly accurate, but the numbers show they should be out there."

More of Everything

Project META and its successor Project BETA (you guessed it: the Billion-Channel Extraterrestrial Assay) ran from 1985 to 1999. Needless to say, neither META–BETA nor SERENDIP nor HRMS ever found candidate signals that could be replicated on a second look. But they set the pattern for the modern age of SETI, from 1995 to the present. It can be described as the era of "more of everything"— except, of course, more US government funding.

More philanthropy After the cancellation of HRMS, the SETI Institute took possession of the detecting equipment it had built, and Tarter hit the road to raise money for a privately funded version of the search. Millions came in from computer-industry tycoons such as the Hewlett-Packard cofounders William Hewlett and David Packard, the Intel cofounder Gordon Moore, and the Microsoft cofounder Paul Allen. Project Phoenix got under way in 1995, using rented telescope time at the Parkes Observatory in Australia, then at Green Bank, and then at Arecibo.

But it was clear that the Institute's work needed its own permanent home, which sparked the idea for a new array with a large number of small-diameter antennas, which together can do the work of one extremely large telescope. The SETI Institute's Allen Telescope Array, made up of 42 separate 6-meter dishes, went into operation at Berkeley's Hat Creek Observatory in 2007. It's named for Paul Allen, who put more than $30 million into the project before his death in 2018.

In 2015, in a potentially game-changing development for SETI another tech-industry mogul, the Israeli Russian entrepreneur and investor Yuri Milner, pledged $100 million over 10 years for a project called Breakthrough Listen. Administered by the Berkeley SETI Research Center (also the home of SERENDIP), the program buys observing time on the new 100-meter Green Bank telescope in the Northern Hemisphere, the Parkes radio telescope in the Southern Hemisphere, and the Automated Planet Finder, an optical telescope at the Lick Observatory near San Jose, California. Together, these instruments collect as much data in one day as previous SETI projects did in a year, Milner has said.[21] Data from Green Bank flows into SETI@home, a program that allows volunteers to donate idle time on their home computers to help sort through the noisy data for weak signals. (This 20-year-old experiment in citizen science and distributed computing is still going strong: you can join in at setiathome.berkeley.edu.)

More frequencies The 21-centimeter, 1,420-MHz hydrogen line isn't the only place on the radio dial where extraterrestrial civilizations might choose to broadcast—it is arguably just the most obvious one because hydrogen is the most common element in the universe. Over the years, SETI scientists have also tried listening at other frequencies, such as multiples of 1,420 MHz. In the movie *Contact*, the alien signal is found at "hydrogen times pi." "Hydrogen times *e*" and "hydrogen times the square root of 2" are equally plausible.

Then, not far away from the hydrogen line are the radiation lines for interstellar hydroxyl (OH) ions at 1,612–1,720 MHz. Barney Oliver called the section of the radio spectrum around the hydrogen and hydroxyl lines the "water hole" because H and OH can combine to make H_2O.[22] A water hole is also a gathering place for multiple species, so the optimistic name stuck.

By and large, though, today's SETI researchers don't have to make as many assumptions about which frequencies aliens might use. Modern detectors built on the same principles as Horowitz's Project BETA spectrometer can listen on billions of narrow channels simultaneously. And even newer technologies may help SETI researchers get a peek around the "microwave window," the narrow range of frequencies (from 1,000 to 10,000 MHz) that can glide right through Earth's nitrogen-oxygen-water atmosphere. Until recently, we couldn't search outside those frequency bands because radio energy at frequencies higher than 10,000 MHz gets absorbed by the atmosphere, and energy at frequencies lower than 1,000 MHz tends to be obscured by galactic background noise. But the Low-Frequency Array (LOFAR) telescope, an array of 44 antennas spread across the Netherlands, France, Germany, Sweden, and the United Kingdom, was built to study stars so far away that the Doppler effect, in this case a consequence of the expansion of the universe, has red-shifted their radiation down to the 10 to 240 MHz range. That's the same frequency range we use for radio and TV broadcasts, so it's been suggested that LOFAR might be sensitive enough to eavesdrop on communications leaked by nearby civilizations.[23] Meanwhile, the Square Kilometer Array, a proposed $2.3 billion project to scatter hundreds of radio telescopes across South Africa and western Australia, would be so sensitive that it could detect Earth-style radio or TV signals leaking from a planet dozens of light years away.[24]

More parts of the electromagnetic spectrum   So far I've discussed only radio SETI. But SETI's association with radio astronomy is a historical accident, tied up with the World War II–era history recounted at the beginning of this chapter. Distant civilizations wouldn't be aware of our radiocentrism, however, and, for all we know, there may be far better ways to send messages between the stars, such as laser light.

Optical SETI has been around as an idea ever since the invention of the laser in the 1960s, but it wasn't put into serious practice until the late 1990s. Paul Horowitz at Harvard reasoned that with existing technology humans could generate brief laser pulses that would outshine the sun by a factor of 1,000 and stay coherent across vast distances. To detect such a signal coming from another civilization, you would need a photodetector capable of recording pulses as short as a nanosecond. (Almost nothing natural generates such high-frequency light pulses, so any signal at that bitrate would likely be artificial.) So that's what he and his team built with funding from the Planetary Society and the SETI Institute. Beginning in 1998, Horowitz's Optical SETI project piggybacked on existing optical telescopes at Harvard and Princeton. In 2006, the team switched over to the new purpose-built All-Sky Optical SETI Telescope (OSETI), designed to scan the entire northern sky every 200 clear nights.

Please note that I just mentioned that "almost nothing natural generates such high-frequency light." Horowitz's OSETI project was vulnerable to regular false alarms, caused mainly by Cherenkov radiation—the quick flashes of light generated when cosmic rays hit particles of gas in Earth's atmosphere. Around 2010, a team led by Frank Drake decided to tackle this problem by installing an optical SETI system with three separate photodetectors at the Nickel Telescope at California's Lick Observatory. By rejecting simultaneous, isolated flashes, the new system reduced the false-alarm rate to one per year. In 2015, also at Lick, the SETI Institute worked with Shelley Wright, a physicist at the University of California at San Diego, to start scanning for laser signals at near-infrared wavelengths. Infrared may well be a better choice for laser-based communication than visible wavelengths because sending an infrared beam takes less energy, and it can pass more easily through interstellar gas and dust.[25]

More parts of the sky   An optical or radio telescope in the Northern Hemisphere can see only the northern sky, but a signal from extraterrestrials is equally likely to come from the southern sky. Part of the story of SETI over the past three decades has been about expansion to telescopes in the Southern Hemisphere. For Project META II in the early 1990s, astronomers took copies of Horowitz's hardware to the Tidbinbala and Parkes observatories in Australia and the Instituto Argentino de Radioastronomía in Argentina. Project Phoenix observed 200 stars in the southern sky in 1995, and SETI work has continued at southern observatories ever since. That said, there's still a great deal of catch-up work to do: out of the 103 radio SETI projects since 1960—according to the SETI Institute's comprehensive Technosearch database—fewer than 20 involved southern observatories.[26]

More types of stars   A star's biography—the stages of its evolution and its overall longevity—is predetermined by its mass. Most targeted SETI searches have singled out medium-size stars like our sun because they evolve predictably over 10 billion years, long enough (in theory) for life to arise on their planets. Researchers have mostly ignored very large stars, which burn out too fast, as well as very small stars such as red dwarfs and brown dwarfs, which last a very long time but don't put out much heat. For these small stars, the "Goldilocks Zone"—the band where planets don't boil or freeze but get just enough energy to keep their water in a liquid state—would be very narrow.

But the reality, as we're learning, is that red dwarf stars and their planets far outnumber stars like our sun and *their* planets. In fact, in 2016 researchers found a planet slightly larger than Earth in the habitable zone around Proxima Centauri, a red dwarf that is our closest neighbor in space, a little more than 4 light years away, and in 2017 they found another orbiting a red dwarf, Ross 128, just 11 light years away. (Those distances aren't as small as they sound. It would still take 75,000 years for the fastest object ever launched, the Voyager 1 probe, to get to Proxima Centauri. And in point of fact, Voyager is going in a different direction.)

The inner rocky planets of dwarf stars would be strange to us—they would be so close to their stars that they would be tidally locked, so that one side always roasts and the other freezes—but recent research indicates they might still be habitable. And because their stars are much older and evolve more slowly, civilizations on these planets would have had more time to develop. "This may be one instance in which older is better," says the SETI Institute senior astronomer Seth Shostak.[27] Since 2016, he has been using most of the institute's time on the Allen Telescope Array for a targeted search of 20,000 red dwarfs.

More nations   Practical SETI efforts were born in the United States and the Soviet Union. Over the decades, Australia, Argentina, and a variety of European nations have pitched in. But one of the most exciting potential advances for SETI is happening in the Guizhou Province of southwestern China. When the government-funded Five-Hundred-Meter Aperture Spherical Telescope, or FAST, was completed there in 2016, it eclipsed Arecibo as the world's largest radio telescope. The size and shape of the dish means FAST can see more of the sky than Arecibo and can be three times more sensitive. FAST has already discovered dozens of new pulsars (fast-spinning neutron stars that emit beams of radiation), and its owner, the National Astronomical Observatories of China, signed an agreement with the Breakthrough Listen project in 2016 to share observing plans, search methods, and data.[28] As of this writing, SETI observations using FAST hadn't yet begun.

Glass Half Full

It's been 60 years since Frank Drake first turned the Tatel Telescope at Green Bank toward Tau Ceti and Epsilon Eridani. The philosophical and theoretical debates about extraterrestrials go on, as they should. But at least these days, despite the ridicule coming from small-minded legislators, we're actually looking for aliens, not just talking about them. Not a night goes by when that search doesn't continue.

The overriding fact, however, is that we have found nothing. The Great Silence continues. There are at least three ways to look at this: glass half full, glass half empty, and what I call "wrong glass."

Let's start with the glass-half-full interpretation. To understand it, think of an actual glass. Go to a beach and dip the glass into the ocean. Did you catch a fish in in the glass? No? You could conclude that fish don't exist. Or you could keep looking.

That's the metaphor Jill Tarter offered back in 2010, when she calculated that the portion of the galaxy that SETI researchers have scanned so far is like a glass of water compared to Earth's oceans.[29] Scientists redid the calculation in 2018 and concluded that it was more like a hot tub's worth of water.[30] In either case, a null result isn't all that surprising. Having caught nothing so far, we could give up on our fishing expedition and walk away from the beach. Or we could grab more glasses, dip them deeper, and keep preparing for the day when we'll become authentic ichthyologists.

Classical SETI *hasn't* failed, from this point of view, because it has barely begun. This attitude is understandable,

The Great Silence
continues.

given that SETI research in the Morrison-Drake-Sagan-Horowitz-Tarter tradition grew up alongside radio astronomy. Undeniably, radio and optical techniques adapted from mainstream astronomy have provided scientists with their first hard, empirical data about extraterrestrials, even if those data are negative so far. To paraphrase Donald Rumsfeld again, you go to war with the telescopes and spectrometers you have, not the ones you wish you had. And given the scope and variety of the universe, there will always be new places to point our telescopes and new frequencies to search.

The glass-half-empty interpretation is more somber. Having dipped our glass locally, we now know—even more firmly than Fermi did—that there aren't any aliens begging to be discovered in our immediate neighborhood. We have examined most of the nearby stars on all the parts of the spectrum that can be measured easily. If there were any civilizations within shouting distance, and if they wanted to be found, we would have stumbled across them by now.

We have not. That must tell us *something* important about the frequency, distribution, or communicativeness of advanced societies in the galaxy. There can't be 50 million star systems inhabited by gossipy civilizations—the upper end of the Dolphins' first estimate. N must be much lower, and it may be exactly 1. SETI has failed so far, according to the glass-half-empty interpretation, because it's a search for something that probably isn't there.

But there are yet other possibilities. Maybe a glass isn't a good tool for catching a fish. Maybe the fish know we're fishing, and they're scurrying out of the way. Maybe we've been looking so hard for something resembling a fish that we've missed the real quarry—say, transdimensional cyborg heptapods.

All of which is to say that there could be something fundamentally wrong or short-sighted about the way we've been doing SETI so far. In the history of science, important problems have rarely yielded to the very first hypothesis. So perhaps it's time to develop a better one about extraterrestrials.

We'll come back to that idea and its potential for resolving the Fermi Paradox in chapter 5. But first we need to review recent discoveries in disciplines that barely existed when SETI began that have changed the landscape of possible solutions.

EXTREMOPHILES AND EXOPLANETS

Scientists have never been of one mind or one mood about the possibility of extraterrestrial life. Huygens's casually open-minded attitude reigned until hard-headed Whewell came along and demanded some real evidence. Schiaparelli and Lowell offered an appealing fantasy about our next-door neighbors, and though the canals of Mars proved illusory, the idea of Martians percolated through popular culture for the next half-century. But by the 1960s, as biologists began to unpack the mechanisms of genetics, the pendulum had swung back the other way. The idea that systems as intricate and improbable as RNA and DNA could pop up by chance on more than one planet had come to seem "so disreputable it verged on the crackpot," as the physicist, writer, and SETI advocate Paul Davies recalls.[1]

Yet there are always some scientists for whom the "crackpot" label isn't a deterrent. Remember that the

1960s was also the time of Project Ozma and the Order of the Dolphin. And today, as Davies notes, the Dolphins' cockamamie optimism rules again. The silence from the skies so far shows that *intelligent* life is going to be an elusive prize, but most researchers believe that we're closer to discovering indications of off-world life of some kind—probably simple life, at first—than at any point in history.

To be clear, we don't have the kinds of concrete evidence we would like. We know from fossils that abiogenesis, the emergence of self-replicating organisms from their nonliving chemical precursors, occurred here on Earth at least 3.8 billion years ago—almost immediately after the planet's hot molten surface cooled and oceans emerged. But so far we have no direct evidence of life on other worlds. A fossil on Mars would be nice, or perhaps a glimpse of a tentacle through an ice crack on Europa or even just the chemical traces of organic processes in a distant atmosphere. We have had no such luck.

But we *are* learning that habitable worlds are far more numerous than we previously supposed and that Earth life is far more versatile than we ever expected. Both are indirect indications that there's a path for the emergence of intelligent life on other worlds. So when Davies writes that "the change in sentiment is due ... to fashion rather than discovery,"[2] I think he's being just a little bit of a sourpuss.

In fact, for the past four decades a revolution has been under way in the young field of astrobiology, fueled by revelations on both the "astro" side and the "bio" side. (By the way, how cool is it that people get paid to be *astrobiologists*? If I were a college student today, this would be the major for me.)

New Life

When Drake and his peers set out on their search 60 years ago, they thought that electromagnetic signals from intelligent extraterrestrials would be our only sure sign that life had emerged elsewhere. But since that time our understanding of how life works on Earth—where organisms live, how they evolve, and what they can do—has shifted radically. Now it seems just as likely that we'll discover evidence of microbial life first, drawing on a new understanding of the types of environments where life can thrive and how life itself alters those environments.

Consider the tale of Carl Woese. A microbiologist at the University of Illinois in Urbana, Woese decided in 1969 that he wanted to figure out when different varieties of bacteria diverged from each other on the evolutionary tree—or what was then thought to be a tree. After spending years analyzing the RNA sequences of ribosomes—a Herculean feat in the days before the advent of automated gene sequencers—Woese published a paper in 1977 announcing that he had discovered an entirely new domain of life, the Archaea. These tiny single-celled organisms had always been mistaken for bacteria, but Woese showed that in fact their genetic code is as different from that of bacteria as it is from our own. On top of that, he discovered massive evidence of gene transfer between archaea: the sideways movement of genes from organism to organism rather than from parent to child and often across species lines. It eventually became clear that the classic Darwinian tree is more like a crazy banyan, with branches that constantly fuse and cross. The biology community resisted Woese's findings for many years, but today we know that archaea are one of the dominant forms of life on the planet and that horizontal gene transfer is commonplace. Indeed, 7 percent of the human genome consists of viral genes inserted by retroviruses.[3]

What's so humbling about this story is that up to the late 1970s there was a whole domain of organisms that we hadn't noticed and that was using a form of genetic variation we hadn't dreamed about. Woese's insights got biologists thinking again about the earliest days of life on Earth and greatly complicated their picture of evolution. If we missed an abundant type of life on our own planet, the find suggested, perhaps our imaginings about life elsewhere were equally limited.

But that was just the beginning: the year 1977 would turn out to be somewhat of an annus mirabilis for scientists interested in what life might be like on other worlds.

The year's next shocker arrived 2 kilometers below the surface of the Pacific Ocean, about 280 kilometers northeast of the Galápagos Islands. Researchers from the Woods Hole Oceanographic Institution (WHOI) were mapping the sea floor using a remote camera when something unlikely turned up on the film: a colony of white clams. That didn't fit with the dogma of the time, which said organisms found at such great depth and in such complete darkness would be rare and alone. So geologists John Corliss and John Edmond boarded *Alvin*, the famous research submarine operated by WHOI, to take a first-hand look. What they discovered near a spring where hot water was seeping up through the seafloor would spark an entirely new science. "It was a fantastic undersea garden, an oasis vibrant with life," as the science writer David Toomey explains in his excellent book *Weird Life*. There were anemones, giant clams, crabs, fish, and mussels, all living in water containing high levels of hydrogen sulfide. WHOI scientists would soon determine that bacteria were chemosynthetic—feasting on the hydrogen sulfide and forming the base of a whole food chain that operated without the benefit of photosynthesis.[4]

Two years later Corliss and Edmond would visit an even stranger undersea location near the Gulf of California, where natural chimneys were pumping out superheated water thick with iron sulfide. These "black smokers" confirmed a theory that hydrothermal vents would be found at the boundaries of Earth's tectonic plates. The unexpected part was the sheer volume of life around the vents. Despite the incredibly high water temperatures—300°C or more at the smokers—the vents were surrounded by mats of chemosynthetic bacteria and, as it turned out, archaea. These microorganisms supported a rich ecosystems of gastropods, bivalves, crustaceans, and annelids, including alien-looking giant tube worms.

The only reason superheated water at that depth doesn't boil is that it's under enormous pressure from the ocean above. Before the discoveries, biologists hadn't supposed that life could thrive amid such pressures and temperatures or such a stew of dissolved minerals and heavy metals, not to mention the lack of sunlight. But the findings touched off an explosion of interest in "extremophiles," organisms adapted to survive in extreme environments that would destroy their more fragile cousins. Research going back to the 1960s on "hyperthermophilic" bacteria growing in 90°C water in hot springs in Yellowstone National Park suddenly seemed more relevant.

As researchers began looking for life in other locations previously considered unlikely, it turned up almost everywhere. "Halophiles"—salt-loving microbes found in the Dead Sea as early as the 1930s—were now joined by "acidophiles" (acid lovers, found in volcanic springs and mines), "psychrophiles" (cold lovers, found in polar ice and permafrost), "barophiles" (pressure lovers, found deep underground or in oceanic trenches), "polyextremophiles" (such as *Thermococcus barophilus*, a sulfur-eating archaeon found in a deep-sea hydrothermal vent), and even "radiophiles" (such as a fungus found converting ionizing radiation into usable energy in the core of the devastated Chernobyl reactor).[5]

It now seems that DNA-based life can go everywhere water can go. In 2013, 4,000 species of bacteria and archaea turned up in samples from a lake in Antarctica that's been sealed off from the sun under 800 meters of ice for at least 120,000 years.[6] In the continental crust, 3 to 6 kilometers below the surface, there's thought to be a "deep biosphere" made up of bacteria and archaea that metabolize chemical food.[7]

The takeaway from all these discoveries is that life is astonishingly adaptable. And that takeaway, in turn, expands astrobiologists' notions about the range of environments that might be habitable on other planets. On Earth, "life is the rule rather than the exception," the US National Research Council observed in 2007 in a report on the basic requirements for life and future directions for astrobiology.[8] There's no reason to suppose that nature is less inventive on other worlds.

It now seems that
DNA-based life can go
everywhere water
can go.

Indeed, it's even possible that there are now Earth microbes on Mars, carried there by NASA's Curiosity rover. The space agency determined after launch that parts of the rover had not been sterilized according to proper procedure, and swabs taken before the launch showed that even the parts of the craft that were properly cleaned with peroxide and ultraviolet radiation harbored more than 60 species of bacteria.[9] It's not known whether these microbes are capable of surviving the long trip to Mars and the frigid conditions on the Martian surface. But from experiments on the International Space Station, scientists know that spores of some *Bacillus* species can stay alive in space—an extreme environment if there ever was one—for at least 18 months.[10]

Viking Invaders

Speaking of the red planet, let's jump back again to 1976–1977. The third big development of that period, from an astrobiologist's perspective, was the arrival on Mars of the twin Viking landers. Viking 1 landed at Chryse Planitia on July 20, 1976, and Viking 2 landed at Utopia Planitia on September 3. To this day, they remain the only probes ever designed and flown with the goal of testing directly for signs of extraterrestrial life.

Both Vikings were equipped with robotic scoops that could retrieve handfuls of Martian soil and dump them into a biological experiment system built to conduct four types of studies. There was a device called a gas chromatograph mass spectrometer that could heat the soil and measure the mass of the vaporized molecules to identify them by their molecular weight. An ingenious "gas-exchange" experiment added organic nutrients to the soil and then watched to see whether it gave off oxygen, carbon dioxide, nitrogen, hydrogen, methane—all potential signs that microbes in the soil might be metabolizing the nutrients. The "labeled-release," or LR, experiment was similar, except that the added nutrients contained ^{14}C, a radioactive isotope of carbon, and the air above the sample was monitored to see if it included radioactive $^{14}CO_2$ or $^{14}CH_4$ gas released from the mixture, either of which would have been another sign that life was at work. A final "pyrolytic-release" experiment exposed the soil sample to a gas resembling Mars's atmosphere, except that the carbon-bearing gases were made with ^{14}C; the idea was to see whether any photosynthetic organisms present in the soil might be converting this radioactive carbon into biomass.

The results from both landers came back negative—for the most part. The gas chromatograph found no traces of organic molecules; the gas-exchange experiment detected no signs of metabolism; and the pyrolytic-release experiment found no evidence of photosynthesis. But the

LR experiment was a puzzling exception. On both Viking 1 and Viking 2, the soil samples absorbed the nutrients and immediately began churning out radioactive $^{14}CO_2$. The effect went away in control samples that had been baked or isolated long enough to kill any microorganisms—which is just what you would expect if there were living, metabolizing organisms in the soil. Given the lack of confirmation from the other experiments, however, NASA downplayed the LR findings. The radioactive carbon detected by the experiment could have been produced by an inorganic oxidation reaction, researchers hypothesized.

But the LR experiment's designer, Gilbert Levin, an engineer at Arizona State University, believes to this day that the landers discovered Martian microorganisms.[11] More than 40 years after Viking, "we are still debating whether there was life on Mars," says Joel Levine, a former NASA research scientist.[12]

Exobiology (as astrobiology was known until the 1990s) is a science with plenty of theory and speculation and not much hard data, and the Viking missions didn't fix that. But this characterization could change when the European Space Agency's Exomars Rosalind Franklin rover touches down on Mars in 2021. The rover will carry an experiment designed to test whether oxidation is a plausible explanation for the Viking LR results. It will also look for microfossils around former bodies of water. And it's equipped with Viking-style spectrometers that will search not just for organic molecules but also for evidence of chirality, or "handedness." The building blocks of life, such as sugars and amino acids, come in "left-handed" and "right-handed" mirror-image versions. For efficiency, Earth organisms prefer to use just one chirality: all sugars are right-handed, and all amino acids are left-handed, for example. If the Exomars rover were to find organic molecules, and if one chirality outweighed the other, it would be a strong sign of life. (Ironically, Levin had planned to include a chirality experiment on the Viking landers, but it was the victim of cost cutting.[13])

Different Seas

There was yet another advance in 1977 that helped change the way we think about life on other worlds. As we saw in chapter 2, that was the year NASA launched the Voyager robotic probes toward Jupiter and the outer solar system. Now that both craft have left the solar system, they're lonely ambassadors whose main remaining mission is to carry their golden records into deep interstellar space. But as they flew past Jupiter and Saturn and their moons between 1979 and 1981, they sent back photos and measurements that jolted planetary scientists into a new realization: there may be abodes for life in our solar system outside of the traditional "Goldilocks Zone" where Earth orbits.

For exobiologists, the excitement focused on three worlds: Jupiter's moon Europa and Saturn's moons Enceladus and Titan.

The photos of Europa from Voyager 1's Jupiter flyby in early 1979 showed only a crust of water ice etched with a network of long scratches, reminiscent of Schiaparelli's *canali* on Mars. But when Voyager 2 zoomed closer to Europa four months later, the stripes showed up as delicate brown veins painted across an incredibly even surface, the smoothest in the solar system. The likeliest explanation: the veins were places where Europa's ice shell had cracked, allowing salty water to well up from an underlying liquid ocean and then to freeze as younger, darker ice.

After the flyby, Voyager scientists speculated that Europa's ocean is kept in a liquid state by tidal flexing, the same process of elongation and friction that powers spectacular volcanoes on Jupiter's innermost moon, Io. NASA's Galileo mission discovered much later that Europa has a magnetic field, supporting the idea that there's a deep conductive layer of salty liquid water. And even later the Hubble Space Telescope spotted huge geysers of water erupting from Europa's south pole.

Life, in any form we can conceive, requires a liquid as a medium for chemical reactions and to shuttle substances such as nutrients and waste across an organism's membrane. Water is ideal in part because of its properties as a solvent. The discovery of a liquid water ocean on Europa—in fact, the largest in the solar system, with two to three times the volume of Earth's oceans—of course led exobiologists to wonder whether life might exist there.

Trapped beneath the icy crust, life on Europa wouldn't get any sunlight, but that's not a showstopper, given what we now know about the chemosynthetic abilities of extremophiles on Earth. In fact, the crust may provide crucial protection from Jupiter's deadly radiation belts. And because astronomers think Europa's oceans interact with its rocky core, it's possible that chemical "food" is abundant down there. (Sometime in the 2020s, a planned NASA mission called Europa Clipper will attempt to fly through the water-vapor plumes, allowing it to sample Europa's ocean for organic matter without ever touching down.[14])

Enceladus, visited by the Voyagers in 1981, turned out to be a mini-Europa, with an ice-covered liquid-water ocean. Its mix of cratered and smooth areas was a sign that parts of the terrain are younger and probably covered with fresh ice from water volcanoes. The source of Enceladus's internal heat remains a mystery because its orbit isn't eccentric enough for tidal forces to play a big part. But the joint NASA–European Space Agency Cassini mission confirmed that there's water under the ice, and samples from plumes revealed a cometlike mix of water vapor, nitrogen, methane, carbon dioxide, and trace hydrocarbons. Cassini also detected molecular hydrogen, a sign that the ocean of Enceladus may be nourished by hydrothermal vents like those here on Earth. No wonder that astrobiologists have put forward at least five proposals for a return to Enceladus to scout for life.

Then there's Titan, Saturn's largest and most intriguing moon. The Voyager probes established that the planet is covered in thick orange smog, produced when the sun's ultraviolet rays break down methane in the upper atmosphere. Underneath the smog layer, the Voyagers found, is a thick and relatively balmy nitrogen–methane atmosphere; the methane produces a greenhouse effect that keeps the surface at −180°C.

That's obviously chilly compared to Earth, but it's warm enough to keep hydrocarbons in a liquid state, suggesting that Titan might have lakes of liquid methane and ethane. Hubble and Cassini later confirmed this hypothesis, revealing Lake Michigan–size pools of the stuff. The Huygens probe, which landed on Titan in 2005, sent back pictures of huge channels carved by methane rivers and a surface littered with water-ice rocks and dusted by dark hydrocarbon snow.

Why should astrobiologists care about lakes of methane? Well, water is the most familiar solvent for supporting life, which is why NASA has long embraced the mantra "follow the water" in its search for life in the solar system. But it isn't the only conceivable solvent. "Our practical search for extraterrestrial life is focused on water-rich planets and moons because of the possibility that they can support Earth-like life," the US National Research Council observed in its report in 2007. However, "that does not preclude other strategies for carbon-based life to thrive in nonaqueous solvents, such as exist on Titan."[15]

No one knows how life might get started in methane. But no one knows how abiogenesis occurred here on Earth, either. Organic solvents such as methane and ethane are generally toxic for Earth's carbon-based microorganisms. But astrobiologists have imagined "cells" with methane-based outer shells protecting molecular machinery that makes water.[16] They have sketched systems of genetic polymers that could keep heritable information intact even in nonaqueous solutions as well as exotic metabolisms based on processes such as the conversion of acetylene to methane. Such forms of alternative or "weird" life have never been observed on Earth, where it seems that abiogenesis occurred just once or that DNA-based life wiped out the alternative forms.[17] But they're not impossible as long as life's basic requirements are present: a fluid environment with an energy source, lots of carbon-containing atoms, and temperatures high enough to allow basic chemical reactions.

Such thinking led the committee behind the National Research Council study in 2007 to use what, for scientists, was some remarkably strong language:

No one knows how life might get started in methane. But no one knows how abiogenesis occurred here on Earth, either.

If life is an intrinsic property of chemical reactivity, life should exist on Titan. Indeed, for life not to exist on Titan, we would have to argue that life is not an intrinsic property of carbon-containing molecules under conditions where they are stable. Rather, we would have to conclude that either life is scarce in these conditions or that [sic] there is something special, and better, about the environment that Earth presents (including its water).[18]

For a scientific culture steeped in Copernicanism, of course, the idea that there's something special about Earth verges on heresy. Which is a big part of the drive to look beyond Europa, Enceladus, and Titan for even more places that may harbor life.

Strange New Worlds

In 1755, the philosopher Immanuel Kant, of all people, published an explanation for the formation of our solar system that turned out to be the correct one: *Allgemeine Naturgeschichte und Theorie des Himmels* (*Universal Natural History and Theory of the Heavens*). In essence, clouds of interstellar hydrogen clump up under gravity into balls, which rotate, collapse, ignite as stars, and form protoplanetary disks. Grains of dust around the stars coalesce into planetesimals, then embryo planets, then (sometimes) rocky planets like Mercury, Venus, Earth, and Mars. If the embryo planets are large enough and far enough away from their star, they start collecting hydrogen and helium driven outward by the new star's energy, and they balloon into giant gas planets like Jupiter, Saturn, Uranus, and Neptune.

Once astronomers figured out the details of this "solar nebular disk model," it seemed obvious that the same thing would happen around other stars. But a remarkably long time went by before they obtained proof that our eight planets aren't the only ones in the galaxy.

The first star to divulge its own planets was a pulsar called PR1257+2, the remnant of high-mass star that exploded in a supernova. By tracking tiny changes in the pulsar's rotation rate, radio astronomers deduced in 1992 that the pulsar has three planets, all of which apparently formed from debris left by the supernova.

But PR1257+2 was an oddball case, and it was overshadowed three years later by the detection of the first extrasolar planet, or exoplanet, around a sunlike star, 51 Pegasi. The planet, later named Dimidium, has a mass at least half that of Jupiter but an orbit that takes it unexpectedly close to 51 Pegasi; this made it the prototype for a class of exoplanets now called "hot Jupiters." The find, made by Michel Mayor and Didier Queloz at the University of Geneva in Switzerland, was the rumble that unleashed an Alpine-scale avalanche of exoplanet discoveries—more than 4,000 at this writing. That avalanche has in turn stirred up a cloud of new ideas about extraterrestrial life and the conditions where we might find it.

Generally speaking, you can't just point an optical telescope at a distant star and see its planets. They're much too faint, and in any case they're obscured by the star's own light. Astronomers usually resort to more indirect methods, and so far they have come up with two very successful ones.[19] The radial-velocity method monitors for any Doppler shift in a star's characteristic spectral lines to look for a wobble as it moves through space—a giveaway that it's being jostled by one or more planets. From the magnitude and period of these swings, astronomers can determine those planets' orbital radii and approximate masses. This is how Mayor and Queloz found Dimidium, which, it turned out, orbits 51 Pegasi in just 4.23 days, at a distance of 0.05 astronomical units (AU), much closer to its star than Mercury is to our sun. But it was just the first planet discovered using the radial-velocity method: as of this writing, 760 more exoplanets have been spotted this way.[20]

The other main technique, the transit method, has revealed many more planets—3,114 so far. This method measures a star's light to see if it dims slightly when one of its planets "transits"—that is, passes between the star and our telescopes. Obviously, this approach works only if the geometry is just right and we happen to be located in the same plane as the star's rotation and its planets' orbits. If we were looking down on a star from its north pole, its planets would circle it but never transit. But astronomers using ground-based telescopes have found several hundred planets this way, and space-based exoplanet-hunting telescopes—Europe's CoRoT mission launched in 2006 and NASA's Kepler mission launched in 2009—have found thousands more. Kepler, which kept its telescope pointed at a small patch of the heavens for four years, checking the brightness of more than 150,000 stars every 30 minutes, is credited with detecting 2,662 exoplanets, including many systems with multiple planets.

Two big surprises emerged from all this work. The first was that stars with planets are the rule, not the exception. Exoplanet hunters have learned that the majority of sunlike stars have one or more planets: the average is now thought to be about 1.6 planets per star. So our galaxy, which includes at least 100 billion stars, is home to at least 160 billion planets, and that's not even counting rogue planets that were ejected from their star systems. (Yes, that's a real thing.)

The other surprise was how few planetary systems resemble our own. Astronomers had grown so accustomed to the layout of our solar system—with its cheerful type G star at the middle, surrounded by small, rocky inner planets and more distant gas giants—that they assumed this layout was inevitable. The discovery of Dimidium was the first clue that this might not be the case. Because large planets are easiest to find, most of the worlds discovered in the early years of the exoplanet boom were, in fact, hot Jupiters.

And we know now that the story is even stranger. In 2017, astronomers in Belgium and Switzerland, using the ground-based Transiting Planets and Planetesimals Small Telescope (TRAPPIST), announced they had found a system with seven Earth-size planets, all orbiting very close to a tiny red/brown dwarf star christened TRAPPIST-1. Located 40 light years away, it's like a clockmaker's miniature solar system: the planets are in tight rings 1.6 million to 8.8 million kilometers away from the star, so close that they all would fit well within the orbit of Mercury. If the planets were orbiting our sun at those distances, they would obviously be toasted to a crisp. But TRAPPIST-1 is a Jupiter-size star with just 8 percent of our sun's mass and a feeble 0.05 percent of its luminosity. That means its Goldilocks Zone is much closer to it. At least three of the planets, TRAPPIST-e, TRAPPIST-f, and TRAPPIST-g, happen to be inside this zone, meaning there's a possibility their surfaces have liquid water. (Alas, if they're inhabited, no one there is signaling us at the moment: the SETI Institute scanned the area of TRAPPIST-1 in 2016 with no results.[21])

From an astrobiological perspective, the big takeaway from this golden era of exoplanet discovery is that there's now a concrete list of potentially habitable worlds other than Earth. If you define "habitable" conservatively—counting only exoplanets that are rocky, with a radius less than 1.6 that of Earth and a mass of less than six Earths, in orbits where they can maintain surface liquid water—then you can point to 19 worlds so far, according to scientists at the University of Puerto Rico at Arecibo. The closest one is Proxima Centauri b, 4.22 light years away. And if you define habitability a little more loosely—counting smaller and larger exoplanets in a slightly wider band of orbits—then there are 33 more, for a total of 52.[22]

That number may go up soon. In 2018, NASA launched an MIT-built spacecraft called the Transiting Exoplanet Survey Satellite (TESS), which is designed to detect Earth- or super-Earth-size planets around nearby stars. Compared to Kepler, TESS is monitoring a much wider field of view. It's also measuring stars' brightness at much more frequent intervals and will eventually cover far more of the sky. Mission planners expect to find at least 50 rocky planets over the mission's two-year course. If any of them are particularly interesting—for example, if they seem to have atmospheres—then astronomers will follow up using NASA's James Webb Space Telescope (JWST), due to launch in 2021.

The JWST's main mirror is much larger than that of the Hubble Space Telescope, which will give it greater resolution and sensitivity. And its instruments are built to study infrared wavelengths, where water, methane, and carbon dioxide happen to have strong spectral lines. The hope, therefore, is that by subtracting the spectra of transiting planets from the spectra of their stars, JWST will be able to reveal the composition of exoplanet atmospheres for the first time.

Researchers will be looking hard for "biosignatures," distinctive mixes of gases that might indicate that living organisms are keeping a planet's atmosphere off-kilter. Here on Earth, for example, the 21 percent concentration of oxygen in our atmosphere should be a dead giveaway to alien astrobiologists that there's life on our world. After all, oxygen is such a reactive chemical that there would be little of it in the air if it weren't for the cyanobacteria, phytoplankton, and land plants that keep releasing it as a by-product of photosynthesis.

Two more gases produced by Earth life, methane and nitrous oxide, are also abundant enough in our atmosphere to be detected from space. But the list of *potential* biosignature chemicals in the atmospheres of inhabited worlds is much longer than that. The MIT astronomer Sara Seager, deputy science director of the TESS mission, has compiled a daunting list of 14,000 of them.[23] Any of these substances in unusual concentrations in exoplanet atmospheres could be the strongest sign yet of life on other worlds.

A New Equation for Extraterrestrial Life

What are the chances that astrobiologists and exoplanet hunters will find off-world life in the near future? Seager has come up with an intriguing way to think about that problem. It's a revised and updated version of the Drake Equation, and it goes like this:

$$N = N^* F_Q F_{HZ} F_O F_L F_S$$

In this context, N is not the number of communicative civilizations, but rather the number of planets with detectable biosignature gases.

N^* is the number of stars that will be closely studied by TESS and JWST. It's roughly 30,000.

F_Q is the fraction of those stars that are "quiet" and therefore friendly to life. It's about 20 percent, or 0.2, Seager guesses.

F_{HZ} is the fraction of quiet stars that have rocky planets in their habitable zones. Seager puts it at 15 percent, or 0.15.

F_O is the fraction of those planets that are observable using the transit method. Unfortunately, this fraction is very small due to the geometry problem described earlier. Seager calculates that it's 0.001.

Those terms are the easy ones to define. The last two are more speculative, and they make all the difference. F_L is the fraction of observable planets with life, and Seager optimistically assumes it will be all of them—that is, $F_L = 1$.

Finally, F_S is the fraction of inhabited planets where the spectroscopic signatures of life will be dramatic enough to be detectable. Seager puts it at 0.5.

When you do the math, 30,000*0.2*0.15*0.001*1*0.5 = 2. In other words, after all the work that the TESS team is doing to find candidate Earth-like exoplanets and all the work JWST scientists will do to study their atmospheres, then *if they're very, very lucky,* Seager thinks they can expect to find signs of life on two of them.

To maximize the chances of success, Seager recommends a highly catholic approach to the spectroscopic search. That means JWST shouldn't search just for oxygen, methane, nitrous oxide, and other well-known biosignature gases but for a large collection of them. In one recent presentation about the "Seager–Drake Equation," she used an oft-cited admonition in the US National Research Council's astrobiology report from 2007. "Nothing," the committee wrote, "would be more tragic in the American exploration of space than to encounter alien life and to fail to recognize it."[24]

Of course, Americans aren't the only ones searching for alien life. Astrobiology is now a global pursuit, with potentially global implications. And it has been greatly accelerated by the discoveries in the past few decades about the environmental extremes that life can tolerate and the number of places that fit within those extremes. To find intelligent life, we must first find life—and now we have many more places to look for it.

ANSWERING FERMI

So, where is everybody? What Michael Hart called "Fact A," the apparent absence of extraterrestrials on our world or any nearby world, has not budged. Thanks to astrobiologists, exoplanet hunters, and SETI researchers, we have much more data to think with than Drake and the other members of the Order of the Dolphin had back in 1961—and a heck of a lot more than Fermi had in 1950. But that wealth of data hasn't made the original problem disappear. In many ways, the Fermi Paradox has only grown more acute.

To see why, let's return briefly to the original Drake Equation, which gives us a quick-and-dirty way to sort the known knowns from the known unknowns and see where the biggest gaps remain. (That's all the equation was ever meant to do. Some scholars have been heaping criticism on it lately, saying it's shallow and unscientific,[1] but given its usefulness as a roadmap and an explanatory tool over nearly six decades, that criticism feels a little unfair.) Here it is again:

$$N = R^* f_p n_e f_l f_i f_c L$$

where N is the number of advanced civilizations in the Milky Way galaxy that ought to be able and willing to communicate. Now, one way to resolve the Fermi Paradox is to assert that Fact A will never be dislodged and that we are, in fact, alone in the galaxy. This is the same as saying that $N = 1$.

How realistic is that? Well, from the work of astronomers who study stellar evolution, we know that R^*, the rate of formation of good, life-friendly stars in our galaxy is about 7.5 stars per year.

From the science on exoplanets that has emerged over the past decade or two, we know that most sunlike stars have planets and that many have more than one, so it's safe to say that $f_p = 1$.

From that same exoplanet research, researchers think that $n_e = 0.3$. That is, out of every 10 planets, roughly 3 should be habitable.

Here on Earth, life emerged almost instantly once there was liquid water to support it, and it has proved to be incredibly hardy and adaptable. There is still lots of room for debate on this, but it would not be out of line to estimate that life arises on one out of four habitable planets; f_l = 0.25.

So far, so good. For these first four terms, we aren't just guessing anymore; we have some actual evidence or some strong indications. By the way, when I cotaught a seminar on the Fermi Paradox at MIT in 2018, my colleague Paola Rebusco and I challenged our students to come up with their best group estimates for each term in the Drake Equation. The values I'm using here are the ones they settled on.

From what has been said so far, the product of the first four terms, $R^*f_pn_ef_l$, is (7.5*1*0.3*0.25) = 0.56. If you buy into those estimates and you want to force the Drake Equation to yield an answer of exactly 1, then the product of the other three terms, f_if_cL, must be 1.79 (since 1.79*0.56 = 1). As a reminder, f_i is the fraction of planets where simple life evolves into complex, intelligent life, f_c is the fraction of planets where at least one intelligent species develops interstellar communications technology, and L is the average lifetime of a communicating civilization.

With a single example to work from—Earth—we can only guess about those three terms. Because there are no hard constraints, our MIT students decided to be optimistic and conjecture that L is very long, the idea being that once civilizations get through their technological adolescence, they have enough know-how to stay alive until their home stars become red giants and incinerate their planets. In the case of our sun, that will happen a billion years from now. (Which seems like plenty of time to devise an escape plan and extend our longevity, but we'll put that complication aside for now.) After adjusting slightly for the possibility of other types of cosmic catastrophes, such as a gamma-ray burst that would wipe out all life on a planet, our students decided that a typical high-tech civilization might last half-a-billion years: $L = 5 \times 10^8$ years.

If L is indeed large, then to bring $f_i f_c L$ all the way down to 1.79, $f_i f_c$ must be extremely small: namely, 3.58×10^{-9}. In other words, the fraction of inhabited planets where microbes evolve into radio astronomers must be about 3.58 in a billion or 1 in 279 million.

To illustrate: let's say you could rewind the tape of Earth's history back to the time right before the Cambrian explosion, 540 million years ago, then let it run forward again. If you did that 279 million times, our math so far says you would end up with a high-tech civilization like ours only *once*.

If you accept these odds, the Fermi Paradox isn't a paradox because Fermi was too optimistic in his back-of-the-envelope calculations. The reason aliens haven't visited is that they don't exist.

This feels like a gloomy solution to the paradox, but it's not impossible. Indeed, if you spend any time thinking along these lines, you realize that strict Copernicanism

Indeed, if you spend any time thinking along these lines, you realize that strict Copernicanism breaks down in an evolutionary context.

breaks down in an evolutionary context. There may be nothing special about Earth, and we humans are just another twig on a convoluted evolutionary bush, but there *is* something unique about us. DNA-based life has had many tries at building brainy, crafty creatures, and humans are the only ones who have managed to slip beyond Earth's atmosphere. Imagine that you found some primordial slime on Proxima Centauri b, then ran the clock forward a few billion years: it would be pretty astonishing if a tool-making species emerged the first time. "This view of life is not bleak; it is simply the way things are," writes Matthew Cobb, a zoologist and evolutionary biologist at the University of Manchester. He continues:

> The fact that we have made it this far does not imply that there must also be spacefaring aliens, nor that we are somehow destined to reach the stars. The apparent inevitability of the existence of human civilization is a trick of perspective, a cosmic tautology: we can only wonder about such matters because we are here. Our existence has not been guided by some supernatural force, nor is it written in our genes. We have just been very, very lucky.[2]

Speaking for myself, I *do* feel lucky to be alive and human in this time and place. "We are a way for the cosmos to know itself," as Carl Sagan observed.[3] To have a part in this adventure is an inconceivable privilege.

That said, any solution to the Fermi Paradox that counts *too much* on luck may be open to suspicion. Perhaps there was only one way in 279 million for evolution to produce modern humans on Earth. But there may be many other pathways to intelligence and technology. To accept the idea that $N = 1$, you have to believe that *every other pathway* on *every other planet* has failed *every single time* throughout the 13.5-billion-year history of the galaxy. That's a lot to ask.

It's so much to ask, in fact, that we are obliged to consider additional solutions to the paradox. Everything we're learning from astrobiology suggests that simple life will turn out to be common in the universe, providing billions of separate starting points for the evolution of intelligence. From that perspective, the Fermi Paradox is not just alive but also more relevant than ever.

As we have been learning, the biggest unknowns in the debate about extraterrestrials have to do with the emergence of life and its evolution from simple microbial forms into intelligent, long-lived, technological species—the $f_l f_i f_c L$ section of the Drake Equation. In this chapter, we'll go on a brisk tour of proposed solutions to the Fermi Paradox, organized according to the way they deal with these four unknowns. This will set us up for a final discussion in chapter 5 of the solutions that seem most viable and compelling and of possible ways forward for SETI and for our curious species.

There's a large scientific literature on the Fermi Paradox, and I'm not the first to try to catalog and evaluate all of the answers scholars have proposed. Deeply curious readers will want to consult *If the Universe is Teeming with Aliens ... Where Is Everybody?* by Stephen Webb and *The Great Silence: Science and Philosophy of Fermi's Paradox* by Milan Ćirković. Both were major sources for this chapter.

But I'm organizing my list in a different way that's intended to show how the potential solutions are moving "up the stack" of the Drake Equation, from the brute physical factors to the more complex biological and sociological factors. Nobody argues anymore that R^* or f_p or n_e are tiny: recent science shows they are not. So the remaining arguments for a very small value of N hinge on f_l, f_i, f_c, and L. Clearly, as we develop the ability to detect biosignatures in exoplanet atmospheres, f_l will be the next factor to move from the "known unknowns" column into the "known knowns" column. That will leave open the much hairier questions about what intelligence really is, how it emerges, how intelligent species develop technology, and what they do with it once they have it. All perspectives on these questions are valuable because we need to get a handle on them in order to know what to search for. But my bet is that we won't get definite answers until we meet some actual extraterrestrials.

Throughout this discussion, keep in mind that as we try to resolve the Fermi Paradox, we're really looking for ways to invalidate any one of its premises, so that the logical incompatibility disappears and the world makes sense again. In this case, the seemingly sound premises are:

1. From what we know about astrophysics and biology, there should be lots of extraterrestrial civilizations in our galaxy.

2. By now, these civilizations should have had plenty of time to expand throughout our galaxy, visiting or contacting every single planetary system.

3. Nothing we have seen so far credibly qualifies as a signal or artifact of intelligent extraterrestrial origin.

One or more of these premises must be flawed—we know that for a certainty. So let's look at the various attempts to poke holes in them.

Life Is Rare (f_l Is Small)

For a long time, one of the plausible ways out of the Fermi Paradox was to argue that planets suitable for life—especially small, rocky, metal-rich planets like early Earth—are uncommon. But thanks to data from Kepler and other telescopes, we now know that this isn't true. We have been able to measure the radii of roughly 3,000 of the more than 4,000 known exoplanets and have found that about one-third are rocky worlds with a radius less than twice that of Earth.[4]

What about water? A rocky world without any H_2O wouldn't be a great place for life. For a long time, geologists weren't sure where Earth's water came from and whether the process that brought it here was unique or easily repeatable in other systems. Now they're pretty sure that Earth's water came from asteroids and planetesimals that were around in the early solar system or maybe from comets originating in the Kuiper Belt, a diffuse ring of asteroids outside the orbit of Neptune. Structures similar to the Kuiper Belt have been detected around many other stars. So the idea that rocky inner planets are usually dry isn't a good way out of the paradox.[5]

The next question is how many planets orbit in their stars' habitable zones—that sweet spot where the star's radiation is melting any water ice but not causing all the water to boil away. The wider this zone, the better for life, especially because the zone moves outward as stars age and get hotter and brighter. Computer models developed in the 1970s suggested that these zones were quite narrow and that for many types of stars *no planet* could stay within the zone long enough for life to evolve.[6] But in 2013 researchers using Kepler data asserted that 22 percent of sunlike stars have Earth-size planets in their habitable zones.[7] And as the TRAPPIST-1 system illustrates, even narrow habitable zones can contain multiple planets. So it's likely that billions of planets in our galaxy are in their stars' continuously habitable zones.

So far, then, our solar system seems pretty average, with nothing to explain why complex life would be unique to Earth. Enter the geologist Peter Ward and the astronomer Donald Brownlee, professors at the University of Washington in Seattle. A book they published in 2000, *Rare Earth: Why Complex Life Is Uncommon in the Universe*, sent out shock waves in the astrobiology and SETI communities. It stitched together a careful, credible, and compelling case that we humans wouldn't be here if it weren't for a remarkable, perhaps unrepeatable combination of circumstances.

Earth has plate tectonics A quick geology lesson: convection in the upper mantle, powered by the planet's radioactive interior, sends magma to spreading centers in the ocean crust, pushing the ocean plates ever outward. The edges of the lighter continental plates ride up over these giant conveyor belts as the leading edges of the ocean plates dive back down into the mantle. Volcano-rich subduction zones form at the overlaps. Plate tectonics is important for life because it acts indirectly as a kind

So far, then, our solar system seems pretty average, with nothing to explain why complex life would be unique to Earth.

of thermostat for the atmosphere. Rock weathering, in which calcium combines with heat-trapping carbon dioxide (CO_2) in the air to form calcium carbonate, helps to ensure that the atmosphere never gets too hot. Volcanism at the subduction zones, meanwhile, takes old calcium carbonate on the sea floor and recycles it, spewing out new calcium-rich surface rocks as well as CO_2, ensuring that the atmosphere never gets too cold. On a planet without plate tectonics, Ward and Brownlee argued, a relatively stable climate like the one that allowed complex life to emerge on Earth would be much harder to maintain.[8]

Earth has a large, nearby moon   The Moon stabilizes the tilt of Earth's rotation relative to the plane of the solar system by balancing the pull from the Sun and Jupiter. Without the Moon's calming influence, it's possible that Earth's tilt would vary wildly on a time scale of hundreds of thousands years to millions of years (as Mars's tilt does), causing regular, massive climate disruptions—again, a tough prospect for complex life.

Earth has Jupiter   The gravity of this giant planet, striding the boundary between the inner and outer solar system, may have helped to clear the inner zone of most of the asteroids and planetesimals zooming around in the early solar system. We know how much damage a single asteroid did when it created the 150-kilometer-wide Chicxulub crater 65 million years ago; it was a bad day for the dinosaurs. A rocky inner planet facing a constant on-slaught of such objects would be an inhospitable place for advanced life.

Earth has Mars Our neighboring planet became hospitable for life, with liquid water and a thick protective atmosphere, before Earth did. (Unfortunately, it didn't stay that way.) And we know that big impacts occasionally kick up Martian debris, some of which eventually lands on Earth. It's conceivable that microbial life began on Mars and then migrated here or even that there was regular commerce, with Mars serving as a lifeboat for Earth microbes at one or more points in our solar system's history.[9]

Based on all these factors put together, it appears that complex life emerged on Earth thanks to "a highly fortu-itous set of circumstances that could not be expected to exist commonly on other planets," Ward and Brownlee wrote twenty years ago.[10] In our future travels around the galaxy, they predicted, we might find many worlds that harbor microbes, but none with animals, let alone radio astronomers. Ward and Brownlee didn't set out to solve the Fermi Paradox, but they *were* consciously taking on one of the traditional SETI tenets: that once life appears on a planet, it evolves almost inevitably toward greater complexity. "If the Rare Earth hypothesis is correct, then [SETI] clearly is a futile effort," they said.[11]

Ward and Brownlee's book provoked a vigorous and healthy debate within astrobiology and SETI communities. The counterarguments fall into two main groups. One set takes issue with specific scientific details that feed into the hypothesis. For example, we're not sure how plate tectonics started on Earth, so we can't say how common or uncommon the phenomenon might be on exoplanets. Volcanism at subduction zones might not be the only mechanism for keeping greenhouse-gas levels in check. Also, recent computer simulations suggest that Jupiter isn't a very good shield; in fact, it may send *more* asteroids and comets our way. And so on.

The other set of counterarguments is more philosophical. The Rare Earth interpretation—that you only get complex Earth-style life when conditions are exactly like those on Earth—can be seen as a case of circular or at the very least unimaginative reasoning. The US National Research Council report mentioned in chapter 3 warned against what it called "terracentricity" and urged researchers to "make a conscious effort to broaden our ideas about where life is possible and what forms it might take."[12] The astronomer David J. Darling argued in 2001 that Ward and Brownlee were really just telling a story about *our* world, not offering a hypothesis about *other* worlds. "What matters is not whether there's anything unusual about the Earth; there's going to be something idiosyncratic about every planet in space," Darling wrote. "What matters is whether any of Earth's circumstances are not only unusual but also essential for complex life. So far we've seen nothing to suggest there is."[13]

Or have we? It's easy to pick apart the specifics of the Rare Earth idea, but there are other perspectives that might lead one back to the conclusion that f_l is small or, just as discouraging, that once life gets started in a given location, it doesn't stick around long enough to evolve into complex forms. We know, for example, that life emerged on Earth as soon as conditions allowed, at least 3.8 billion years ago. We have a growing understanding of how life works, but we still can't say how it began. And it seems that this happened only once—at least, we have found no evidence of a "shadow biosphere" of non-DNA-based life on Earth. Until we can reverse-engineer the process from scratch, we have no way of judging how likely or unlikely it is on other planets. So it's possible that abiogenesis here on Earth was an incredible fluke and that no other planet will ever be so lucky.

And there are even more bumps on the road from f_l to f_i. For about 2 billion years after abiogenesis, Earth was ruled by microscopic, single-celled organisms (prokaryotes). Somewhere around 1.2 to 1.6 billion years ago, the first multicelled organisms (eukaryotes) emerged. And then *they* swam around for another billion years doing simple eukaryotic things—eating prokaryotes and each other and producing oxygen. Finally, 540 million years ago there was enough oxygen in the water and the air to support animals with larger, more complex body shapes. Almost all of the interesting stuff in evolution (no disrespect to bacteriologists intended) has happened in that last one-eighth of the planet's history.

One possible solution to the Fermi Paradox, then, is that the leap from prokaryotes to eukaryotes is so hard that it doesn't happen on most planets. This is one scenario included under what's called the Early Great Filter Hypothesis: the idea that one of the steps in evolution, though we don't know exactly which one, is extremely improbable.[14]

Related to this is the idea that our solar system features a special kind of engine for evolution, keeping speciation going or nudging it along at key moments. In 1986, the physicist John Cramer called this phenomenon "the pump of evolution." He was drawing on the then new ideas that (1) evolution seems to proceed in fits and starts, with long periods of stasis in between—the "punctuated-equilibrium" theory most closely associated with the paleontologist Stephen Jay Gould—and (2) catastrophes such as asteroid impacts can cause mass extinctions, opening up ecological niches for new species. Perhaps, Cramer speculated, some periodic alignment of the planets in their orbits knocks objects out of the asteroid belt and sends them raining them down on Earth. This doesn't happen so often that it extinguishes promising species, but it doesn't happen so infrequently that evolution can revert to its usual slow pace. "That's my explanation for the Great Silence: we haven't been contacted by an Elder Race because we *are* the Elder Race," Cramer wrote. "We happen to have evolved on a planet where evolution is pumped to progress faster than almost any other place in the universe."[15]

Intelligence Is Rare (f_i Is Small)

Cramer invoked asteroids and other killers from space as a way to explain why evolution led to intelligence on Earth. Contrarily, other scientists invoke catastrophes to explain why it always fails to do so elsewhere.

The universe is a dangerous place—that much is incontrovertible. Life must contend with planetary-scale disasters such as asteroid impacts, supervolcanoes, global ice ages, and Venus-style runaway greenhouse processes. On top of these planetary adversities, there are interstellar hazards such as supernovae, gamma-ray bursts, and stray black holes. One school of thought is that the latter type of disasters are more frequent in the galaxy than we Earthlings, who have escaped or survived them to date, fully realize. To quote Milan Ćirković's summary of this position, "The evolutionary chain leading to intelligent beings is regularly interrupted, and no technologically advanced species arise except by a freak exception."[16]

There's no evidence-based way to refute such arguments. To judge whether they're a good explanation for the Fermi Paradox, we'll have to watch the skies and refine our estimates of the real risks of apocalyptic events such as gamma-ray bursts. (Gamma-ray bursts are truly horrifying events in which imploding supernovae spray surrounding systems with sterilizing radiation; it's thought that a gamma-ray burst may have caused the Ordovician extinction 450 million years ago. If they are frequent enough, they might explain why we're alone and make us fear for our own future.) To me, the only slightly fishy thing about solutions that invoke catastrophe is that *we are always the freak survivors*. To buy into these explanations is, once again, to accept that there is something very special or lucky about Earth. For someone steeped in Copernicanism, this just feels wrong.

Then again, that feeling could potentially be chalked up to a different form of anthropocentrism that scientists call the "observation selection effect." That's what happens when you let the way you evaluate evidence be swayed by the fact that *you exist* to collect the evidence. Just because humanity made it this far, for example, doesn't mean we can infer that the risks of planetary or galactic catastrophe must be mild. Similarly, just because we're intelligent, we can't infer that intelligence must be common in the universe.

The observation selection effect is an important form of cognitive bias, and it will come up again in this chapter. But it can cut both ways. I would say it's hard at work in the Rare Earth Hypothesis, for example, where Earth is held up as the prototype for all animal-inhabited planets. So as we proceed through more explanations for the Great Silence, we'll need to stay on our guard against all forms of anthropocentric thinking on both the glass-half-empty side and the glass-half-full side.

It is, in fact, rash to assume that intelligence is widespread when we don't even have a great idea of what intelligence is or exactly why we acquired so much of it. Other species, such as earthworms and cyanobacteria, have been around far longer than humans and have spread just as far around the planet without learning to do mathematics or write plays. And that's to say nothing of consciousness, which seems to be an emergent property of complex brains, but which we can't yet explain or simulate and which has an unclear and possibly ornamental relationship with intelligence. (It's been argued that consciousness isn't a prerequisite for survival and that even complex organisms can move "intelligently" through their surroundings without being conscious.)

That's why there's a whole class of explanations for the Fermi Paradox that says intelligence and consciousness are flukes or that there's some kind of filter that prevents organisms from evolving past a certain level of smarts. The strongest argument for this viewpoint may be that even through 540 million years of wild experimentation with animal forms, evolution on Earth has produced only one species that can ask the "Are we alone?" question. (Of course, there were other species in the genus *Homo* who might have asked the question, such as *Homo neanderthalensis* and *Homo denisova*, but it seems that *Homo sapiens* eliminated them through competition and interbreeding.) Maybe it's easy to get to dog-level or even ape-level intelligence, but learning how to talk and thus to transmit learning and culture in groups is an insuperable step. Or maybe intelligence is a fleeting trait: species have it for as long as it provides adaptive value, but then it becomes unnecessary or even maladaptive.[17]

Or maybe many species become intelligent and develop language but never go on to invent science, a systematic way of becoming less wrong about the world. The Scientific Revolution sparked by brave thinkers such as Copernicus looks like a self-evidently sensible and advantageous step in retrospect, but it was far from inevitable. Stephen Greenblatt suggests in *The Swerve* that it might never have happened if Poggio Bracciolini hadn't recovered the manuscript of Lucretius's *De rerum natura*. That's a playful theory, but it does remind us that human history is full of bizarre contingencies. From the perspective of the Cosmic

Calendar—a storytelling device popularized by Carl Sagan that compresses the universe's 13.8-billion-year past into 365 days—the whole age of science since Galileo fits into the last second of December 31. So it may be unwise to assume that sentient species on other planets will have used their time in the same way.

Technology Is Rare (f_c Is Small)

Many Earth species build things (bees, wasps, termites, bowerbirds, beavers) or use objects found in their environments as tools (otters, chimpanzees). Researchers have watched crows shaping sticks into hooks and even assembling multiple parts into a single tool.[18] But only humans cooperate to improve tools over time.

Sometime in the nineteenth century, we hit on a formula for doing that at an accelerating rate, and today exponential technological progress feels almost normal. But historically, of course, it's anything but normal. The Neanderthals used the same basic tool kit of stone blades and hammers for 120,000 years with little innovation along the way. So it's certainly possible to imagine planets where the sentient inhabitants make tools but never become avid engineers or where the improvement is so slow that it never leads to steam power, let alone spaceships or radio communications.

As a technology journalist who believes that our inventions have on the whole been a powerful force for good, I would like to think that intelligence, language, science, and engineering are a package deal: each one makes the next easier to acquire, and once you get started, you're on your way to the stars.

But I can hold this view only because I grew up in an era of astonishing technology triumphs. It's the observation selection effect at work. We have to accept the possibility that no species in the galaxy has advanced further than we have or even that we have come to the top rung of our own ladder and will never become a starfaring culture.

In that spirit, the next explanation for the Fermi Paradox switches gears. It assumes that f_l, f_i, and f_c are large and that technological civilizations can and do arise but that other obstacles put a ceiling on their development.

Technological Civilizations Burn Out (L Is Short)

When Drake, Sagan, and the other Dolphins sat down to debate the values that should be assigned to the terms on Drake's blackboard, it was 1961, near the apex of the Cold War. The Bay of Pigs invasion had failed a few months earlier. The Cuban Missile Crisis was less than a year away. When they got to L, the lifetime of a communicative

Sad to say, in the post–Cold War era we have invented even more ways to snuff ourselves out—or at least to set ourselves back by a few centuries.

civilization, their thoughts naturally turned to nuclear annihilation. Surely any species advanced enough to discover radio would also discover how much energy is locked up in atomic nuclei and how it can be released through a chain reaction of fission or fusion events.

Small wonder that when Sagan later illustrated the Drake Equation in *Cosmos*, his icon for *L* was a mushroom cloud. Here's how he thought about the problem:

> It is hardly out of the question that we might destroy ourselves tomorrow. Suppose this were to be a typical case, and the destruction so complete that no other technical civilization—of the human or any other species—were able to emerge in the five or so billion years remaining before the sun dies. Then $N = N^* f_p n_e f_i f_i f_c f_L \approx 10$, and at any given time there would be only a tiny smattering, a handful, a pitiful few technological civilizations in the galaxy. ... Civilizations would take billions of years of tortuous evolution to arise, and then snuff themselves out in an instant of unforgivable neglect.[19]

Sad to say, in the post–Cold War era we have invented even more ways to snuff ourselves out—or at least to set ourselves back by a few centuries. In addition to the still-present threat of nuclear war, there's the danger of uncontrolled pandemics, facilitated by jet travel and antivaccination movements; of deliberate bioterrorism or biowarfare; and of the cumulative self-induced punishments of climate change, including drought, wildfires, superstorms, and sea-level rise. On the more exotic side, some thinkers worry that we'll be destroyed by superintelligent machines or nanotechnology run amok or a particle-physics experiment that rips a hole in space.

And those are just the threats we can imagine in our near future. What might happen to civilizations once they start venturing into the galaxy? One solution for the Fermi Paradox says galactic societies always collapse because they can't find new territory or new resources fast enough to stay ahead of population growth.[20] Another says they're hemmed in by disease or degenerative medical conditions and wind up using all of their resources on health care.[21] Yet another says that they die of boredom after they run out of new things to investigate.[22] And another supposes that they build self-replicating robotic probes to assist with exploration and colonization, but these probes malfunction and wipe out their makers and everyone else.[23]

All of these ideas would make entertaining storylines for science-fiction novels, but to me they reflect our own current-day anxieties more than any predictable laws of galactic empire building. One solution that does seem more persuasive is ecological: it's the idea that the Milky Way's "climate" evolves over time, like a planet's. In the early history of the galaxy, according to this idea, gamma-ray bursts and other catastrophic events may have been more common than they are today, periodically wiping out emerging life forms across many star systems. Only now is the galaxy calming down, meaning that there are fewer catastrophes and fewer resets. We don't see any other civilizations yet because they (and we) are just getting going. With luck, we'll have time to figure out how to protect ourselves before the next blast.[24]

At the University of Oxford in England, an entire organization, the Future of Humanity Institute, is devoted to the study of these kinds of existential risks. It was founded by the philosopher Nick Bostrom, who not only is a prominent $N = 1$ partisan but has also published an entire book about the observation selection effect.[25] Bostrom has written that he hopes we never discover microbes on Mars or any other planet. To him, such a discovery would imply that life is common, but because it appears that civilizations are not common, we have to conclude that most civilizations do destroy themselves or get destroyed.[26]

Bostrom's argument is, in turn, a version of an idea called the Great Filter, first proposed by economist Robin Hanson in 1998. Because space looks empty and dead, Hanson wrote, there must exist "a great filter between death and expanding lasting life"—an extremely hazardous moment through which any advanced species must pass. For humanity, the all-important question is whether that filter is safely behind us in our evolutionary history or still looms ahead. From a Great Filter perspective, it would be a very bad thing to find any evidence of extraterrestrial life because it would be a sign that evolution is straightforward (in other words, that f_l, f_i, and f_c are large) and that our biggest existential challenges still lie ahead (that L is shorter than we would like). "The easier it was for life to evolve to our stage," Hanson summed up, "the bleaker our future chances probably are."[27]

It's an intriguing argument, but, to my eye, a premature and overly pessimistic one. First off, to believe in the Great Filter scenario, you have to accept that space really is empty and dead: that extraterrestrials are not just apparently absent but actually absent or extremely rare. But we can't say that yet. There are too many known unknowns and unknown unknowns. You also have to believe that the Great Filter is so terrible that no civilization can see it coming and outsmart it. That's a depressing and defeatist idea. I side with the physicist David Deutsch, who observes that "almost no one is creative in fields in which they are pessimistic."[28] Now that we have science—a surefire method for asking better questions, gathering more knowledge, building better explanations, and inventing more powerful ways of controlling the world—it's reasonable to believe we are on unstoppable upward course, gamma-ray bursts and other great filters be damned.

Technological Civilizations Grow Slowly

But perhaps expansion is not inevitable. The next three categories of possible solutions for the Fermi Paradox go at the problem in yet another way. They don't put any limits on the terms of the Drake Equation. Rather, they assume that we *aren't* alone, and they seek to explain why our putative neighbors remain undetected.

One approach is to question one of Fermi's key premises: the idea that the galaxy has been around long enough for at least one civilization to have colonized all of it. When you think about it, empire building is a rather human trait. There might be all sorts of reasons that extraterrestrial societies don't have an insatiable drive to expand.

Maybe they're hermits and don't wish to leave their home planets or communicate with other intelligent beings.[29] Maybe their planet is, like Venus, perpetually cloudy, so they never develop astronomy or space travel.[30] Perhaps they calculate in advance that perpetual growth would be too costly or unsustainable.[31] Or perhaps they do try to cross the galaxy, only to find that they don't have the resources or the willpower to keep going.[32] (Note: I'm breezing through all these explanations, but each one has been proposed by serious researchers and described at length in peer-reviewed journals.)

The slow-expansion concept has some intriguing variations. Maybe every extraterrestrial race builds a virtual-reality matrix and gets so immersed that they have no desire to explore the real world.[33] Maybe they're happy with their telescopes, and they decide they can find out everything they want to know about the universe without traveling.[34] Maybe their societies always slip into totalitarianism, and they lose the creative or capitalistic drive needed for expansion.[35] Maybe the galaxy still includes pockets, voids, or wildernesses that extraterrestrials haven't infiltrated, and our solar system just happens to be in one of them.[36] Recent mathematical simulations suggest that the galaxy could be more or less fully settled but still peppered with zones that go unexplored for millions of years.[37]

Or maybe—to return to the idea Fermi himself proposed to unravel the paradox—there's a decent number of civilizations in the galaxy, but they're just too far apart to make travel or communication worthwhile.

You may be wondering where our MIT students came down in the Drakeology exercise that I mentioned at the beginning of this chapter. As noted, they guessed that $R^* = 7.5$, $f_p = 1$, $n_e = 0.3$, and $f_l = 0.25$. They went on to estimate that $f_i = 0.001$ (i.e., that intelligence emerges on one in a thousand inhabited planets), $f_c = 0.6$ (6 in 10 intelligent species develop advanced technology), and $L = 500$ million years. Plugging in all those values, their final estimate for N was 16,875 communicating civilizations in the Milky Way.

If this is in the right ballpark, it may help to explain why we haven't yet been contacted by extraterrestrials, despite the fact that we have been spreading unintentional calling cards—the terrestrial radio and televisions signals that leak into space—for more than 80 years.

Assume that the galaxy is a very flat cylinder with a radius of 50,000 light years and a height of 1,000 light years. Its volume is thus roughly 7.85×10^{12} cubic light years. If there are 16,875 communicating civilizations inside that volume, and if they are spread out evenly (which is a bit of an oversimplification because the center of the galaxy is probably inhospitable to life), then the average distance between any two communicating civilizations is 870 light years.[38]

We know of a super-Earth-size exoplanet exactly that far away from us, in the constellation Cygnus. It's called Kepler-1229b, and it's a rocky world with about 1.4 times Earth's radius, orbiting within the habitable zone of its star, a red dwarf.[39] If there were a civilization on Kepler-1229b, and if that civilization has a telescope that can resolve events on Earth, then at this moment in 2020 its populace would be looking at us in the year 1150, just after the Second Crusade ended in fiasco and humiliation for Europe's Christian kings. Our earliest TV signals would not reach them until the year 2806, and even if they responded immediately, we would not hear back from them until 3676.

In other words, *space is big*—"vastly, hugely, mind-bogglingly big," to quote Douglas Adams.[40] Unless the number of communicative civilizations in the galaxy is in the high tens of thousands, or they have set up lots of automated listening posts, there may not have been time for our signals to reach them. What physicists call our "future light cone" simply hasn't extended very far. In a low-density galaxy, these other civilizations wouldn't know that we're here or that we're busy building radios and rockets and wondering about aliens.

(Of course, the size of Earth's future light cone does not limit *incoming* signals. Whether we can pick up radio or optical signals from other civilizations depends only on how far away they are and when they started transmitting. For example, if there's an inhabited star system 3,000 light years away but civilizations in that system began broadcasting only 2,000 years ago, their messages wouldn't have reached us yet.)

Technological Civilizations Are Uncommunicative

Another way out of the paradox is to suppose that the number of civilizations capable of communicating with us is large but that the number who *choose* to communicate with us is small.

Unless the number of communicative civilizations in the galaxy is in the high tens of thousands, or they have set up lots of automated listening posts, there may not have been time for our signals to reach them.

Perhaps, for example, there are plenty of aliens, but they all have left their original star systems in the galaxy's interior to hang out somewhere safer and more peaceful, such as the galactic rim. Or maybe they're nomads, traveling around in world ships that they refuel using interstellar gas clouds; if they have no use for planets, they wouldn't be interested in us.[41]

One of the most disquieting ideas along these lines is that most civilizations stay silent because they know there are malevolent forces at work in the galaxy, and so they don't dare call attention to themselves.[42] Scientists are paying more attention to this possibility now, thanks to a new wave of proposals to send messages into space to short-circuit our SETI searches and provoke aliens into answering. (As mentioned in chapter 2, this process is called METI, messaging extraterrestrial intelligence. It's also sometimes called "active SETI.")

To opponents of METI, the Great Silence may be a sign that terror reigns in the galaxy. Just in case, they say, we should probably avoid giving away our own existence. "The consequences could be catastrophic, as any civilization detecting our presence is likely to be technologically very advanced, and may not be disposed to treat us nicely," the *Nature Physics* columnist Mark Buchanan opined in 2016.[43] Indeed, some scientists fear that any kind of contact with extraterrestrials could quickly go sour. "If aliens visit us, the outcome would be much as when Columbus landed in America, which didn't turn out well for the Native Americans," the late cosmologist Stephen Hawking warned in a documentary in 2010.[44]

It's hard to say which side is guiltier of giving in to human biases, fears, and hopes: those who assume aliens would be vicious, imperialistic monsters (see the films *The War of the Worlds* [1953], *Alien* [1979], *Independence Day* [1996]) or those who assume they would show godlike benevolence (*2001: A Space Odyssey*, *Close Encounters of the Third Kind* [1977], *Contact*). Rather than try to psychoanalyze beings we have not yet met, we should have a calm and rational discussion about the merits of METI while continuing our existing listening programs—an activity that now needs the retronym *passive SETI*.

Meanwhile, there are related explanations for the Great Silence that don't involve as much projection. Perhaps extraterrestrial civilizations know about us and just aren't interested enough to reach out. After all, if they have spotted us, they *are* likely to be technologically very advanced, so what could we possibly have to offer them? They might appreciate Sylvia Plath, Prince, and eggplant parmigiana, but they aren't likely to need sheetrock or Segways.

Then there's the Zoo Hypothesis, sometimes called the Interdict Hypothesis. It's one of the oldest and most famous solutions for the Fermi Paradox, and it was first set down in scientific detail by the MIT radio astronomer John Ball in 1973. It's an argument by analogy. "We do not always exert the power we possess," Ball wrote. "Occasionally we set aside wilderness areas, sanctuaries, or zoos in which other species are allowed to develop naturally." If extraterrestrials feel that we Earthlings need more time to "develop" before we're ready to enter into meaningful relations with them, they would probably cordon us off as completely as possible because, as Ball observed, "The perfect zoo ... would be one in which the fauna inside do not interact with, and are unaware of, their zookeepers."[45]

Science-fiction authors floated similar ideas well before Ball's paper appeared. *2001: A Space Odyssey*, which was adapted from Arthur C. Clarke's novel *The Sentinel* (1951), imagined that Earth had zookeepers who left a key to the front gate in the form of the monolith buried on the Moon. The Prime Directive, first mentioned in an episode of *Star Trek* in 1966, says Starfleet personnel can't interfere with the social development of less-advanced cultures (that is, until they develop warp drive, and then all bets are off).

It's a convenient idea. It says the Great Silence is so silent because it's engineered that way. It's even flattering, in a way, because it says an advanced civilization thinks we're worth preserving in our uncontaminated form. After all, we're doing the same when we try to seal off the uncontacted tribes of the Colombian Amazon or North Sentinel Island.

But the idea's convenience is also its weakness. If the theoretical extraterrestrials are intent on hiding from us, then it will never be possible to prove or disprove the hypothesis until the moment we're released from the zoo. That's bad news in modern science, where the principle of "falsifiability," introduced by philosopher Karl Popper in 1959, is widely embraced. The principle says that in order to be considered scientific, a theory has to come with a built-in way to show that it might be wrong. Unfalsifiable hypotheses are typically shunned because there's no way to design experiments around them. (That said, there are ideas, such as string theory, that have large scientific followings despite being untestable given our current technology.)

On top of the falsifiability problem, the Zoo Hypothesis requires that *every* society and *every* individual in the supposed Galactic Club cooperate to keep the quarantine in place. That's a great deal of trouble for them to go to, given the vastness of our past light cone. Ideas like this violate the nonexclusivity principle, which says that we should be skeptical toward scenarios that require a high level of conformity to work when diversity is the usual rule in nature.[46] (The same criticism applies to many of the explanations described so far.)

Wild Solutions

All of the Fermi Paradox solutions I have listed so far have their good points and bad points. In the next and final chapter, I describe a class of answers that are more persuasive to me than the rest. (Yes, I'm saving the best for last.) But before moving on, for completeness's sake I want to mention a few more random ideas. All of them have been advanced as possible explanations for the paradox, but, to my mind, they're weird, wild, improbable, wildly improbable, or just plain dumb. Judge for yourself.

Flying saucers This solution says that Fermi's premise that aliens haven't visited us is wrong. If they're already here and buzzing around in (rarely spotted) UFOs, then there is no paradox. Because it is seemingly impossible to write about SETI without addressing this issue, let me say it for the record: just because a flying object goes unidentified doesn't mean that it is piloted by aliens. There isn't a single example of a UFO sighting or an alleged alien visitation or abduction for which an unbiased scientist would resort to an extraterrestrial explanation sooner than a terrestrial or psychological one.

Alien artifacts This is similar to the flying-saucers idea, except it says that aliens visited thousands of years ago and may have helped to build the pyramids or to scratch out the Nazca geoglyphs in Peru. Among many other shortcomings, this idea is an insult to our ancestors

There isn't a single example of a UFO sighting or an alleged alien visitation or abduction for which an unbiased scientist would resort to an extraterrestrial explanation sooner than a terrestrial or psychological one.

and their art and architecture skills. But I do agree that we should stay on the lookout for objects of extrasolar origin elsewhere in the solar system.

The strong anthropic principle The idea here is that the universe was made for us and only us. Not by a deity, of course—rather, all of the important parameters and constants in physics seem to be fine-tuned to allow or, indeed, to require the emergence of humans as conscious observers. Basically, this principle is the observation selection effect—which is usually a trap—twisted around and made into an explanation.

The Planetarium Hypothesis This insidious variation on the Zoo Hypothesis says that the galaxy is teeming with visible alien activity, but we're unaware of it because the universe we see is a projection or illusion engineered to keep us in the dark. It's not clear whether the originator of this idea, the English hard-science-fiction author Stephen Baxter, meant it seriously. Even he agrees that the illusion couldn't be maintained for long once humans start venturing into space in earnest because "a perfect simulation would exceed the capacities of any conceivable virtual-reality generator."[47]

The Simulation Hypothesis What if it isn't the sky that's the simulation—what if it's *us*? The idea that our entire reality is a virtual-reality sim built by superintelligent extraterrestrials has been advanced by Nick Bostrom, among others, and it has caught on lately with figures such as Elon Musk.[48] To me, this hypothesis is yet another product of our own contemporary obsessions and anxieties and an example of science-fiction tropes trickling down into daily life. The 1930s had *The War of the Worlds*; we have *The Matrix* (1999).

The Doomsday argument In 1993, the Princeton astrophysicist J. Richard Gott published a paper in *Nature* declaring with 95 percent confidence that the human race will go extinct sometime between 5,000 years from now and 8 million years from now.[49] He didn't have a crystal ball or any special insight into evolution. He based the prediction on the seemingly Copernican idea that if you come across a thing at a random moment in time, it's unlikely you are seeing it at the very beginning or the very end of its lifetime. In fact, you can easily calculate that because you probably didn't show up in the first quarter or last quarter of the thing's lifetime, there is a 50 percent chance that it will last another one-third to three times its current age. This calculation allowed Gott to predict in 1969 that there was a 50 percent chance that the Berlin Wall would be gone by 1993, and, in fact, as we know, it came down in 1989. The same math says that there's a 95 percent chance that an observed thing (humanity, in the case of the *Nature* paper) will last another $\frac{1}{39}$ to 39 times its present age, hence his 5,000- to 8-million-year estimate. And applying the same reasoning to SETI and alien populations, Gott concluded that there can't be any galaxy-wide civilizations or even many civilizations with populations larger than ours because if there were, you (the random observer) would probably be a member of one of them. The math is simple and hard to argue with, but it's also hard to swallow. For one thing, we don't know if our position as observers is random or not.[50]

Living in our own little bubble   This idea arises from inflation theory in cosmology, which predicts that our patch of space–time is a "bubble universe" born as a tiny fluctuation in a vacuum and that there's an infinite number of other bubble universes that are unreachable from our universe. If this "multiverse" idea is true, then there would be an infinite number of universes with exactly one advanced civilization each, an infinite but smaller number of universes with two or more advanced civilizations each, and an infinite but much, much larger number of universes with no advanced civilizations. From the multiverse perspective, the odds that we happen to exist in one of the bubble universes that has more than one advanced civilization are basically zero—we're much more likely to be in a universe with exactly one civilization (us). More math trickery? Perhaps.[51]

Panspermia   I have already mentioned the idea that Earth life started on Mars. That idea is a limited form of panspermia, the theory that abiogenesis occurred somewhere other than Earth and that life arrived here after traveling through space, perhaps under the care of an ancient race of planet seeders. It's a trivial solution to the Fermi Paradox in the sense that it declares "there are aliens, and they are us." But it doesn't say anything about the chances that there are beings on other worlds. It doesn't even answer the question of how life began—it just pushes it back a step.

Black holes   In this scenario, we don't see any extraterrestrial civilizations because they're all living near or even inside black holes. In these regions, space, time, energy, and matter are compressed, theoretically making computation and other processes faster and easier.[52]

Transcendence   This solution says we don't see extraterrestrials because they have evolved into a form we can't perceive or understand. They might be postbiological, having uploaded their minds to their computers, singularity style. In effect, they would be like gods. It's a concept cloaked in mysticism. Aliens that advanced would probably be uninterested in us and incomprehensible *to* us, so I'm not sure we should even include them in the group of beings we're trying to detect. Adjacent to transcendence is the proposal from the Polish science-fiction giant Stanislaw Lem that very old civilizations blend into the background, literally. Their technology becomes indistinguishable from nature, and "we perceive it as operations of the laws of physics," as Ćirković puts it.[53] It's a mind-blowing idea, but I don't know how we would ever test it, unless we stumbled across some clear sign of artificiality inside reality, like a diagram of a perfect circle built into the digits of pi (an idea that comes up at the very end of the novel version of *Contact*).

The menagerie of ideas in this chapter is, if nothing else, a display of human imagination. We can think of many ways to be alone.

If SETI continues to come up empty, one of these explanations may eventually emerge as the most powerful. Or a few of them may coalesce: it may be the case that intelligence is rarer than Fermi thought, *and* interstellar expansion is just as hard as he thought.

The only easy way out of the problem would be to pick up an extraterrestrial signal or identify an extraterrestrial artifact. Then we would face a different set of questions, about whether and how to establish two-way communications. But at least our isolation would be over.

I hope it's clear, at this point in the book, that the biggest uncertainties about the existence of extraterrestrials are no longer the physical ones about whether there are enough habitable places for other intelligent beings to evolve or whether life is sufficiently tough and creative to spread across the universe. Astronomers and astrobiologists have filled in big parts of that picture in the past two to three decades. It may be too soon to dismiss the Rare Earth Hypothesis, but all of the signs are hopeful. The remaining explanations for the Great Silence bunch up around squishier matters such as how cognition evolves; what it takes to become an outward-facing species that can act on its interstellar dreams; what political, economic, or ecological barriers might stand in the way of a society's growth over the very long term; and how we might communicate with beings who are very different from us.

If it stretches our imaginations to ponder how other advanced societies might deal with these problems, it's because we haven't solved them for ourselves.

JOINING THE CONVERSATION

Science loves an anomaly—a piece of data that doesn't fit with the dominant theory. Focusing on that anomaly can be an opportunity to (*a*) confirm the dominant theory by discovering something new that accounts for the anomaly but still fits within the theory or (*b*) revise or discard the theory and come up with something different that fits the old data *and* the new data. Either way, glory and prizes await. But sometimes scientists get stuck on *a*—doing backflips to defend the existing theory—and forget about *b* because they're not ready to think that far.

In the 1840s and 1850s, nature granted astronomers two delightful anomalies. Neither Uranus nor Mercury was precisely where it was supposed to be according to the predictions of Newtonian orbital mechanics.

The problem with Uranus was that at certain points in its orbit it seemed to be running ahead of the position predicted by astronomical tables but then later would be running behind that position. In 1845, the English astronomer John Couch Adams and the French mathematician Urbain Le Verrier calculated independently that there must be another planet beyond Uranus whose pull was creating a slight boost when Uranus was catching up with it and a slight drag after it had passed.

Le Verrier's math proved to be more accurate than Adams's. The German astronomer Johann Gottfried Galle pointed his telescope in the direction Le Verrier instructed, and, voilà, there was Neptune. For discovering the new planet "with the point of his pen," in the words of his colleague François Arago, Le Verrier was awarded the Royal Astronomical Society's Copley Medal in 1846.

Flush from this victory, Le Verrier tackled Mercury. The innermost planet has a highly elliptical orbit, and the point of its closest approach to the sun—the perihelion—advances or "precesses" slightly with each swing, so that over time Mercury's orbit traces out a kind of daisy petal or Spirograph pattern. This pattern fits with Newtonian dynamics. The anomalous part was that Mercury's orbit precesses twice as fast as Newton's formulas predict it would. In 1859, Le Verrier, applying the same ideas he used to find Neptune, hypothesized that there must be an undiscovered planet inside Mercury's orbit, subtly tugging Mercury forward. Sight unseen, he named that planet Vulcan, for the Roman god of fire.

Le Verrier's paper touched off a race to find the theoretical new planet. But Vulcan turned out to be less cooperative than Neptune. Over the next four decades, numerous astronomers watched for objects transiting the sun's surface or peeking out beside it during total solar eclipses. False alarms abounded, but no observation was ever confirmed.[1]

Finally, in 1915 Albert Einstein published his general theory of relativity. It overturned the old Newtonian understanding of gravity and in the process offered predictions that precisely agreed with Mercury's observed precession without requiring any new planets. The difference between the Newtonian and Einsteinian calculations, it turned out, is the curved fabric of space–time, which is dragged around by the mass of the sun even as Mercury and the other planets move through that fabric. Le Verrier was brilliant—but he was no Einstein.

As we think about extraterrestrials and how to find them, it would be smart to keep the story of Vulcan in mind as a warning.

The anomaly, in the case of SETI, is the Fermi Paradox. The aliens are not here even though theory predicts they ought to be.

To resolve the anomaly, the SETI community has turned to their most favored and familiar tools, radio and optical telescopes. They have tuned them to search

Le Verrier was brilliant—
but he was no Einstein.

for signals on the frequencies they think extraterrestrials would choose if they wanted to get our attention.

By this point, they have spent more time listening to these frequencies than nineteenth-century astronomers spent searching for Vulcan, and the strategy has been no more successful. There have been a few "Wow!" moments, but no alien signal has been confirmed on a second look.

What should we do now? Conclude that extraterrestrials are as imaginary as Vulcan was? Surely not. Keep sampling the ocean one glass at a time? Look for a new method? Or reconsider the dominant theory?

The answer must be some combination of these options, infused with a new humility about our limited perspective and an awareness of what the National Research Council astrobiology committee called our "terracentricity."

As this short book winds down, I examine the possibility that *we're not looking at the problem the right way.* Just as Einstein had to go beyond Newton to explain Mercury's precession, we might need to come up with new and better ideas about how extraterrestrials think before we'll be able to sync up with them.

On the one hand, Jill Tarter's argument that it's too early to give up on SETI when we have examined only a fraction of all stars on a fraction of possible frequencies is incontestable. Unless the galaxy is saturated with civilizations, which it clearly isn't, we may just have to search for a while. The fact that passive SETI continues at all in the United States long after Congress blocked government funding for the search is a testament to the superhuman patience and resourcefulness of people such as Drake, Tarter, Shostak, and Horowitz.

On the other hand, it isn't selfish to hope that the search will actually succeed in our own lifetimes. Even passionate admirers of SETI must consider the possibility that the whole enterprise is being held back by our current technological capacities and ways of thinking. Like the drunkard searching for his lost keys under the streetlight, we have been looking only where we know how to look.

What if the signs of extraterrestrials are all around us, and we've been failing to recognize them? That's what I called at the end of chapter 2 the "wrong-glass" idea. It's also the premise of a final group of Fermi Paradox solutions. I have saved them for last because I think they're the most provocative and possibly the most productive.

Signal and Noise

One variation on the wrong-glass idea is the possibility that the radio spectrum is the right place to listen for deliberate messages from extraterrestrials, but astronomers fixated too early on supposedly "magic" channels such as the hydrogen line. Maybe there are interesting frequencies we haven't considered. Maybe we should be looking at higher or lower frequencies or figuring out ways to scan trillions of channels across the entire radio spectrum. Or maybe the aliens aren't into channels at all, and they use something analogous to ultra-wideband technology, in which extremely short pulses are transmitted across a huge spread of frequencies.

Another interpretation is that we're asking too much of our hypothetical pen pals. For instance, maybe they don't know that our atmosphere absorbs radio waves and that only a certain range of frequencies can slide through it, so they have no particular reason to transmit within the terrestrial "microwave window." Or maybe they're transmitting at much lower power than we might hope.

In either case, we might need to think about building larger telescopes or bigger and more distributed arrays to increase the baseline for interferometry. Or we might need to take drastic steps to minimize background noise—for example, by putting radio telescopes in orbit or on the far side of the moon. That said, just building better equipment smacks of "more of the same," which hasn't worked so far.

Another idea for finding extraterrestrials is to forget about beacons and instead eavesdrop—to try to catch them communicating with each other, the same way any alien civilization that picks up our own radio or television leakage would be doing. But unless alien broadcasters are being as wasteful as we are, it's doubtful whether we would perceive their signals as artificial.[2] "Physicists have shown," Stephen Webb observes, "that if a message is sent electromagnetically and has been encoded for optimal efficiency, then an observer who is ignorant of the coding scheme will find the message indistinguishable from blackbody radiation."[3] Blackbody radiation is the heat given off by all objects at a temperature higher than absolute zero. If the extraterrestrials are into this kind of efficiency, then we'll have no hope of overhearing them.

A more general statement of this problem is that we aren't sure we'll recognize signals from extraterrestrials *as* signals. Researchers working in the traditional SETI mold have always assumed that a radio transmission from aliens would take a screamingly obvious form, such as a sequence of prime numbers broadcast at the 1,420 MHz hydrogen line or a multiple of it. But maybe the aliens aren't going for obvious. Or maybe their idea of obvious is still a century or two beyond our math.

Along those same lines, it's possible that extraterrestrial civilizations are broadcasting using modes we might not even recognize as communications media. As an extreme example, take gravitational waves. For an entire century after Einstein predicted gravitational waves, physicists weren't certain that they existed or, if they did, whether they could be detected. Then in 2015 the Laser

We aren't sure we'll recognize signals from extraterrestrials *as* signals.

Interferometer Gravitational-Wave Observatory (LIGO) experiment picked up the splash in space–time from the merger of two black holes in a distant galaxy—a discovery that has opened up a new era of gravitational-wave astronomy.

I attended a talk at Harvard University in 2018 where Marek Abramowicz, an astrophysicist from Göteborg University, suggested that one unmistakable way for an advanced alien civilization to signal its presence would be to build a sort of gravity-wave beacon. His proposal: create an artificial black hole with the mass of Jupiter and put it into the innermost stable circular orbit around Sagittarius A*, the supermassive black hole at the center of our galaxy. Abramowicz calculated that a LIGO-like detector would be capable of sensing the gravitational waves produced by this arrangement. If the orbit of the smaller black hole did not decay over a period of one to five years—implying the aliens had found some energy source to keep it from falling into Sagittarius A*—we would know for sure that the object was artificial.[4]

As far as I know, LIGO isn't yet being used for this kind of SETI, and I think Abramowicz's talk was intended as a provocation rather than a serious suggestion. The takeaway, for me, was that there may be more methods for communicating than are dreamed of in our philosophy. Maybe neutrinos, quantum-entangled particles, or tachyons could be used to encode and carry information across interstellar distances. No one can say yet—and those are just the *known* unknowns.

Message in a Bottle

Another article of faith in traditional SETI has been that aliens are unlikely to come here physically when it's so much cheaper and faster to communicate using radio waves or laser light. And, in a way, that assumption feels more natural than ever. In the Internet age, we seem to move around less and less, and we're quickly abandoning physical media such as CDs, DVDs, and Blu-ray discs in favor of ubiquitous broadband and streaming.

However, throw a DVD-RAM disk across the room, and you have just achieved a data transfer rate 250 times that of home broadband.[5] This idea is what network researchers call "sneakernet," and it might not be a bad way to send messages between stars. And sometimes there is no substitute for physically going to a place or at least sending an automated proxy. We had to send a dozen astronauts to the surface of the Moon to find the rocks that nailed down the giant-impact theory for the Moon's formation. And long before astronauts ever set foot on Mars, NASA unraveled deep mysteries about the planet's history by sending the Sojourner, Spirit, Opportunity, and Curiosity rovers rambling across its once watery surface.

What if aliens aren't bound by our assumptions about economics? Might they build physical probes that fan out to many potentially inhabited systems?[6] If they did, would we even know what to look for? Perhaps not. In fact, it's possible that one buzzed right by us in 2017, and we didn't even notice until it was too late.

I'm talking about 'Oumuamua, the skyscraper-size visitor from another star system that passed inside Mercury's orbit in early September 2017 and then flew within 33 million kilometers of Earth (about 85 times the distance to the Moon) around October 7.

Astronomers noticed 'Oumuamua 12 days after its Earth flyby as it was already heading back into deep space. It was picked up by the Panoramic Survey Telescope and Rapid Response System (Pan-STARRS), an installation in Hawaii built to detect asteroids that might collide with Earth.

Interestingly, 'Oumuamua did not behave like an asteroid. Avi Loeb, the chair of Harvard's Astronomy Department, drew attention to its odd qualities in a scientific paper and blog post in November 2018.[7]

First off, Loeb argued, 'Oumuamua should not exist at all. If it is an asteroid, and it arrived through random chance, then there's something drastically wrong with our models for how large asteroids get ejected from their home systems and how abundant they are in interstellar space. Second, the object's motion indicated that it was at rest relative to the average circulation of local stars around the Milky Way. In other words, we plowed into 'Oumuamua, not the other way around—as if it had been placed in space like a buoy. Third, judging from the way 'Oumuamua's reflected light waxed and waned as the object tumbled, it must be extremely flat and elongated and made of something much shinier than rocks and ice. Finally, as it swung around the sun, it picked up more speed than it should have from gravity assist alone. Comets sometimes get extra speed from outgassing, but 'Oumuamua didn't seem to be giving off any gas.

All of these anomalies put together led Loeb to speculate that the extra speed was coming from solar-radiation pressure—the momentum imparted by photons from the sun—and that 'Oumuamua is, in fact, an artificial light sail. He suggested that it may have been forgotten as debris or possibly left for us to find.

A light sail is just what it sounds like: a kite made from a very large, thin mirror that pulls a payload through space using the pressure of starlight or laser light. We have already used the technology: Japan's Interplanetary Kite-Craft Accelerated by Radiation of the Sun (IKAROS) had a 14-by-14-meter solar sail that tugged it to Venus in 2010. The Breakthrough Starshot initiative, backed by Yuri Milner, proposes to use light sails driven by Earth-based lasers to accelerate tiny "nanocraft" to one-fifth the speed of light, allowing them to reach Alpha Centauri in just

Loeb [speculated] ... that ‘Oumuamua is, in fact, an artificial light sail. He suggested that it may have been forgotten as debris or possibly left for us to find.

20 years. Loeb, who is an adviser to the Starshot project, points out that light sails would also be an efficient way to move larger cargo between planets or between stars—so it might not be surprising to find an old one drifting in space.

Loeb's paper drew criticism on Twitter from colleagues who felt there could be simpler explanations for 'Oumuamua's behavior. But in press interviews Loeb was unapologetic:

> To me, not even putting [aliens] on the table for discussion is a crime. Because if you look at the history of science, Galileo Galilei argued that the Earth moves around the sun and he was put under house arrest for that. Now, this of course didn't change the facts. It doesn't matter what is being said on Twitter. This thing is what it is, right? … I don't see extraterrestrials as more speculative than dark matter or extra dimensions. I think it's the other way around.[8]

By the time 'Oumuamua was detected, it was too far away for astronomers to snap photos that would have settled the matter. But it might be possible to catch up with the curious object. A British nonprofit called the Initiative for Interstellar Studies is assessing the feasibility of a mission that would take as little as five years to reach 'Oumuamua using a Jupiter flyby trajectory. The challenge with this plan would be slowing down the craft once it gets there so that it can study the object at leisure.[9]

To Loeb, the important thing is to be ready for the next visitor. He points out that the Large Synoptic Survey Telescope in Chile, scheduled to be go online in 2023, will be far more sensitive than Pan-STARRS and should be able to see 'Oumuamua-like objects coming well in advance of their arrival in our system. "Certainly we will see many more objects that originate outside the solar system," Loeb told *Haaretz* magazine. "Then we'll find out whether 'Oumuamua is an anomaly or not. ... It's possible that space is filled with sails like these and we just don't see them."[10]

A Very Wide Sea

Let's be clear: chances are that 'Oumuamua is a natural object. If we send a probe to visit it, at the very least we'll learn something new about asteroids, comets, and their quirky behavior. If Loeb's brave conjecture turns out to be right, it'll be one of the greatest scientific discoveries of all time. The point is that it's the investigation of anomalies that pushes us forward.

And having stuck with me this far, dear reader, you deserve to know what I really think about the anomaly at the center of this book: the Fermi Paradox. Why is it that we seem to be alone? It can't be because we *are* alone—the galaxy is too vast for that, and once life gets started, it is too inventive and tenacious. My favorite explanation is a mashup of several of the ideas we have encountered throughout this book.

First off, SETI's negative finding so far is itself a kind of result. It tells us that there probably aren't any galaxy-spanning civilizations (what Kardashev called Type III) and that there aren't any civilizations within a few tens of light years of us who care to respond in some obvious way to the signal flares we have been sending up for 80 years in the form of radio and television broadcasts.

So if there *are* extraterrestrial civilizations out there, then either (A) they are located too far away for our broadcasts to have reached them, or (B) they aren't interested in sending back a message that we can recognize, or (C) they are both far away *and* talking in a way we can't understand.

If there is anything to my scenario A and the nearest inhabited systems are simply too far away for our news to have reached them, then that's an argument for a sparsely inhabited galaxy—one with tens or hundreds of extraterrestrial civilizations, but certainly not millions. This explanation feels plausible. Complex life does seem special, and technology even more special; *how* special, we can't say. And the stars are very far apart, making commerce and communications difficult at best.

Yet if there is even one long-lived civilization elsewhere in Milky Way, we still have to ask Fermi's original questions: Why *hasn't* that civilization spread out across the whole galaxy? Why don't we see its members or their equipment?

Here I fall back on biology, economics, and physics. Our own mortal, fleshy, planet-bound bodies aren't made for long space journeys. To leave our solar system, we would need to build heavily radiation-shielded generation ships or develop some form of suspended animation or engineer a new strain of hardier humans or perhaps upload our minds to machines. Why should we expect any other civilization to undertake such an arduous and expensive journey unless it were monumentally important? Even if extraterrestrials decided to bother with in-person travel, it's hard to imagine why our solar system would be a compelling destination for them. In any case, we haven't been advertising our presence long enough for anyone to reach us at sublight speeds (which are the only ones possible).

Don't misinterpret me: I think that if we can survive this century and establish a foothold in places such as the Moon, Mars, the asteroid belt, and Europa, we will eventually become interstellar explorers. But in any foreseeable future our trips to other star systems will have to be surgical and precise. (Beyond the foreseeable future, of course, predictions are meaningless.) Expansion for expansion's

I think that if we can survive this century and establish a foothold in places such as the Moon, Mars, the asteroid belt, and Europa, we will eventually become interstellar explorers.

sake—to capture territory, resources, or slave labor—is, one hopes, a marker of our colonialist past on a finite globe. To expect it of other civilizations feels like a particularly narrow-minded form of anthropocentrism.

Data gathering is a better reason to travel. And if science is the mission, it's far easier to send robots. So why don't we see other civilizations' automated craft? Well, until very recently we didn't have the technology to spot such craft. 'Oumuamua, an object the size of the Empire State Building, nearly slipped out of our solar system unnoticed. If we hadn't built a telescope in the 2000s to search for threatening near-earth asteroids, we would have missed it altogether. And when we barely have the ability to detect suspicious objects in our backyard, it's no surprise that we have failed to see signs of engineering elsewhere in the galaxy.

So this is the first part of my favorite explanation: the galaxy is a very wide sea, and the archipelago of civilizations is sparse. Even if we build signal fires on our highest peaks, it's possible that we will barely see one another. Voyages between islands will be difficult and rare. Any given island might be so isolated that its inhabitants will despair of ever contacting anyone.

But I think my scenario B is equally likely. It builds off the idea that our current approach to SETI is incomplete: there's something about the situation we're just not getting.

Recall the Zoo Hypothesis mentioned in chapter 4. It's the idea that we live in a kind of wildlife sanctuary that's been quarantined by the other civilizations around us, perhaps for our protection, perhaps for theirs. John Ball, the originator of the hypothesis, didn't explore all of the metaphor's implications in his paper in 1973. But an obvious one is that in the zookeepers' judgment we haven't evolved to the point where we can be allowed to communicate with the people outside the zoo; such communication would presumably be either unsafe or pointless. We might one day qualify for release from the zoo if we were to reach some unspecified level of social organization, scientific insight, or technological advancement—but who knows where this threshold might be.

As I noted earlier, one of the major weaknesses of the Zoo Hypothesis is that it would be almost impossible to maintain a zoo or quiet zone large enough to enclose us—meaning not just our solar system but also all the radio sources we can measure from here. And even if it were possible, why would anyone go to so much trouble just to keep us in the dark for a few more centuries or millennia?

But on top of those logistical objections, the Zoo Hypothesis is unacceptably sad and disheartening. To believe it is to buy into the idea that we're not worthy to join the larger community of civilizations and that we're being silently judged according to rules we aren't allowed to know. With apologies to Jordan Peele, it's as if the Galactic Club were keeping all of humanity in the sunken place. If that's what's going on, maybe it's not a club we should want to join.

For all these reasons, I think we need to put aside the Zoo Hypothesis, but it does lead to a related idea that might be more persuasive and comforting.

On the Porch

Perhaps on summer evenings when you were a child, your parents took you along to neighborhood parties where all the kids played tag or chased fireflies in the backyard while the adults sat on the porch, sipping drinks and talking.

You loved to play in the deepening twilight. But, being a curious nine- or ten-year-old, you also wondered what deep, important, and forbidden subjects the grownups might be talking about.

You wandered by the porch occasionally. But whenever you interjected or asked a question, the adults would laugh kindly and say—though not in so many words—*you'll understand someday.*

Then 10 or 20 years went by, and you found yourself going back to such parties, but now you sat on the porch and watched the kids out in the yard. And you realized that grownups don't talk about the future of humanity or the nature of good and evil. They talk about school board elections and TV shows and office politics.

So here's the idea. What if extraterrestrial civilizations are out there, conducting their own business, having their own long conversations, doing nothing special to disguise themselves, but also doing nothing special to translate for our benefit, and we simply don't have the tools to understand?

As noted in the previous sections, they may be communicating with technologies or media we haven't discovered. They may be encoding their messages in ways that don't look artificial to us. They may be exchanging robotic trading ships that are too small for us to spot. They may think or talk in a nonlinear way that we would not recognize as language. In each case, they would remain entirely invisible to us until we developed a way to perceive them—rather the way atoms remained unseen from the time of Democritus right up to the moment J. J. Thomson finally discovered the electron in 1897.

Since Thomson's day, we have developed many ways to modulate electromagnetic signals—for example, by varying their amplitude, frequency, phase, polarization, and time of arrival. Perhaps there are more ways. Let's say, for instance, that in the year 2077 we discover a way to transmit massive amounts of information using quantum-entangled radio-frequency photons. Maybe the moment we turn on our new quantum radios, we'll get an instant download of the Declaration of Sentient Rights and the entire Encyclopedia Galactica.

At that point, by definition, we would be joining the conversation on the porch. "Ah, there you are," our neighbors from other worlds might say. "We wondered when you'd come along."

In other words, maybe it's not the extraterrestrials' job to admit us to their club. Maybe it's our job to unlock the door of our own imaginary zoo and walk out.

Life as We Do Not Know It

My scenario A (the aliens are too far away) and scenario B (they're talking above our heads) aren't mutually exclusive. In scenario C, both are true. Civilizations in our galaxy are too far apart to visit one another or us, but they're still doing their best to communicate, just in a way we haven't discerned.

I am not an astronomer or an astrobiologist, and my proposed solutions to the Fermi Paradox are not necessarily meant to stand alongside others that have been debated, published, and peer reviewed in the scientific literature. And they score low on the falsifiability scale; there's no way to test them short of inventing some new technologies and contacting some actual aliens. That said, my solutions do conform to the nonexclusivity principle, which counsels that we avoid explanations requiring great conformity across space, time, or cultures. And they try to sidestep anthropocentric assumptions about what extraterrestrials would "naturally" think or do.

As we look back at SETI's 60-year history and consider what to try next, our biggest obstacle may be ourselves. We grew up on a planet where the environment shaped life and life shaped the environment in historically contingent and surely unrepeatable ways. When SETI began in the 1960s, it was an outgrowth of developments in just one field—radio astronomy—that was shaped by accidents of commerce and war. The first SETI researchers zeroed in on the frequency 1,420 MHz at a time when the ink was barely dry on Ewen, Purcell, and Oort's discovery of the hydrogen line. Being pragmatic folks, these astronomers started looking for the kinds of signals they hoped alien civilizations would send if they were trying to minimize the detection burden on us. Of course, those signals were the same kinds that radio astronomers were confident they could find; at the time, this conclusion was a conscious and necessary act of anthropocentrism.

But here's the thing: SETI has persisted in the style pioneered in the 1960s even as our capabilities and understanding have grown. We have learned that life can take weird and surprising forms, and we have discovered potentially habitable locations that don't resemble Earth. Worlds with very different histories will be home

Worlds with very different histories will be home to very different beings, so perhaps it's time to reformulate SETI to account for a much broader range of possible targets.

to very different beings, so perhaps it's time to reformulate SETI to account for a much broader range of possible targets.

That is the argument laid out by Nathalie Cabrol, who became director of the SETI Institute's Carl Sagan Center for the Study of Life in the Universe in 2015. Radio and optical astronomy, Cabrol wrote in the journal *Astrobiology* in 2016, are "only focused toward testing one very specific, largely anthropocentric, hypothesis about extraterrestrial intelligence, when data increasingly suggest that there are probably as many distinct life-forms and intelligences as habitable planetary environments in the Universe." We have been searching all this time for "other versions of ourselves," Cabrol argued, when we should really be searching for "life as we do not know it." She urged that, ultimately, "SETI's vision should no longer be constrained by whether ET has technology, resembles us, or thinks like us."[11]

What would a broader vision for SETI look like? Probably much more like astrobiology, where researchers have worked hard to identify universal "signatures" of life that we might potentially detect from Earth. One example, mentioned in chapter 3, would be a systemic disequilibrium in an exoplanet atmosphere in the form of gases such as oxygen that probably wouldn't be there unless they were constantly replenished by living organisms. Alongside these biosignatures, there might be "technosignatures":

tipoffs that an alien civilization has been modifying its planet's atmosphere on large scale (as we are certainly doing here).

Listening strategies might also need to change. We don't know what kinds of neural or sensing systems extraterrestrials might have, Cabrol points out, so we can't say how they would organize their perceptions or thoughts or what kind of alphabet they might use to frame them. Perhaps mathematics is the universal language, as SETI researchers have always assumed—or perhaps it only seems that way to us. There might be other equally fundamental ways to share ideas.

Whatever the case, Cabrol argues, we need to cast a wider net, reconsider what we mean by intelligence, and challenge ourselves to think more like aliens. SETI began as an offshoot of astronomy, but now it needs to become a project engaging all of the sciences. "For SETI," she writes, "it is critical to fully embrace the multidisciplinary approach that was scripted 50 years ago in the Drake Equation and create a well-stocked and diverse tool kit."[12]

Preparing for Contact

If we were finally to succeed in our search for extraterrestrial intelligence, then what? In this book, I have deliberately avoided exploring postcontact scenarios. There are plenty of great stories, books, movies, and TV shows that do exactly that.[13] The details of a real-life SETI success would depend so much on the form or content of the exchange that it's impractical to game it out in any rigorous way.

I'll just say this: I don't think we need to fear that an announcement about contact with aliens would cause panic in the streets or some kind of global nervous breakdown. To be sure, it would forever alter our view of our place in the universe. But Copernicus's and Darwin's ideas were equally revolutionary and destabilizing in their time, and they didn't stop blacksmiths from smithing, schoolteachers from teaching, or merchants from keeping their books. And as I observed in chapter 1, we have a great cultural storehouse of tales from myth, religion, and literature that prepare us for the idea that there are other beings in the sky.

Indeed, one might argue—and writers such as Carl Jung actually have—that there's a preexisting niche for extraterrestrials or for creatures like them in the human psyche.[14] Perhaps the urge to fill this niche in a secular and scientific age is part of what motivates SETI researchers and their supporters.

So let's not spend too much time speculating about an inherently unknowable event. Here's a different suggestion. Perhaps it would be interesting to use this time before contact to ask ourselves what we Earthlings might contribute to an interstellar society and what we ought to be doing to prepare for that opportunity.

Before we can be useful citizens of the galaxy, we have lots of work to do to get our own house in order. There are hundreds of billions of sentient beings right here on our planet—the other animals—whom we barely understand and whom we treat, for the most part, with abominable cruelty and carelessness. Science, technology, and free markets have melded into an engine of unprecedented wealth and prosperity, but we allow that wealth to be distributed in shockingly unequal ways. We have an increasingly comprehensive picture of the human impact on Earth's climate and ecosystems, but we are deeply reluctant to start managing that impact in a responsible way. We have fought many wars and enacted many laws to stamp out prejudice and hate, but we continue to elect leaders who trade in those emotions. We took our first steps on another world in the 1960s and 1970s, but then we lost our collective will and redefined "space exploration" as traveling to low-earth orbit.

In short, we are an imperfect and inconsistent species. But if we do have off-world neighbors, they, too, may be limited. And we have much to interest them: beautiful art and music, millennia of moving and perceptive literature, the hard-won lessons of our history, our soaring dreams—the very kinds of evidence we chose to place on the Voyager Interstellar Record.

We are still writing our story. Will it remain our own, or will it merge into the larger story of intelligent life in the universe? We or our descendants may learn the answer someday. For now, we can only keep wondering and searching.

GLOSSARY

abiogenesis
The emergence of living organisms from inorganic substances. On Earth, as far as we know, this has occurred only once.

Allen Telescope Array
A group of 42 radio telescopes, each 6.1 meters in diameter, at the University of California–Berkeley Hat Creek Radio Observatory in northern California, used simultaneously for astronomical observations and SETI.

Archaea
A domain of prokaryote microorganisms once thought to be bacteria; first classified as genetically separate from bacteria by Carl Woese and George Fox in 1977.

astrobiology
An interdisciplinary field focused on understanding the possible origins and distribution of life outside Earth and how it might be detected. Formerly known as exobiology.

Cambrian explosion
An event roughly 541 million years ago in which existing multicelled animals diversified into a great number of new, more complex species never before seen in the fossil record.

CETI
Communication with extraterrestrial intelligence. The study of the composition of messages that might be understood by extraterrestrial civilizations and the deliberate transmission of such messages. Also known as METI, messaging extraterrestrial intelligence.

chemosynthesis
The conversion of carbon dioxide or methane into biomass such as amino acids and sugars, used by organisms in low-light regions such as the deep oceans or Earth's crust as an alternative to photosynthesis.

Copernican principle
The general assumption—growing in spirit from the observation by Copernicus that the sun, not Earth, is at the center of the solar system—that there is nothing special about one's own vantage point.

Drake Equation
A rough quantitative roadmap, first formulated by radio astronomer Frank Drake in 1961, that organizes the major questions bearing on the abundance of communicative technological civilizations in the Milky Way galaxy.

Doppler shift
The tendency of waves in a signal to bunch up or stretch out, depending on whether the source and the recipient are moving toward one another or away from one another.

Encyclopaedia Galactica
The fictional compendium of all knowledge. Used in science-fiction stories by Isaac Asimov and others and as the title of the SETI-focused episode of Carl Sagan's *Cosmos* TV series (1980).

exobiology
The original term for astrobiology. NASA long maintained an Exobiology Program, and it's still an element of the agency's Astrobiology Program, but the term fell out of general use by the 1990s.

exoplanet
An abbreviated form of *extrasolar planet*: any planet in a planetary system outside our own.

extremophile
An archaean or other organism adapted to live and reproduce under extremes of temperature, pressure, darkness, radiation, or chemical concentration that would be deadly for most familiar organisms.

falsifiability
Testability; the idea suggested in 1959 by the philosopher of science Karl Popper that all good scientific hypotheses should be refutable by experience.

Fermi Paradox
The question "Where is everybody?" first formulated in 1950 by physicist Enrico Fermi. It consists of three seemingly sound premises, at least one of which must be wrong: (1) There should be many extraterrestrial civilizations in our galaxy. (2) The galaxy is old enough that at least one of these civilizations should have colonized the whole of it. (3) We do not see any evidence of activity by extraterrestrials.

gamma-ray burst
A short but enormous flash of high-energy gamma rays thought to be released during a supernova; thought to be strong enough to wipe out all life on nearby star systems.

gravitational waves
Ripples in the fabric of space–time produced by accelerating masses. The first direct measurements of gravitational waves occurred in 2015.

Great Silence
One term for the absence of evidence that extraterrestrial civilizations are attempting to communicate with us.

habitable zone
Also known as the Goldilocks Zone: the band around a star in which orbiting planets could have liquid water on their surfaces.

HRMS
The High-Resolution Microwave Survey (previously known as the Microwave Observing Program), a NASA-funded SETI program that aimed to scan 10 million radio frequencies from the Arecibo Observatory in Puerto Rico.

hydrogen line
The spectral line or "bright" spike in the electromagnetic spectrum seen when neutral hydrogen atoms flip between two ground states and radiate photons. This spectral line is at the precise frequency 1,420.405752 MHz, and waves at this frequency can penetrate interstellar clouds and dust, leading the first SETI scientists to propose that it would be a favorable and obvious frequency for artificial transmissions.

Kepler telescope
A now-retired space telescope named after astronomer Johannes Kepler and operated by NASA between 2009 and 2018. The Kepler used the transit method to locate 2,662 exoplanets.

light cone
A concept taken from relativity theory in physics. The light cone for a given observer at point p at time $0 + t$ is the expanding sphere of space (a cone, in space–time terms) that can be reached by a flash of light—or any piece of information traveling at the speed of light—that leaves p at time 0.

materialism
The idea first embraced in writing by ancient Greek and Indian philosophers that matter is all that exists and that mind and consciousness are by-products of material processes.

METI
Messaging extraterrestrial intelligence. See CETI.

microwave window
The window of your microwave oven. No, just kidding. It's the range of radio frequencies roughly from 1,000 MHz to 10,000 MHz considered ideal for interstellar communication in traditional SETI work.

nonexclusivity principle
The idea that, all things being equal, a theory that accommodates diversity is preferable to one that requires great conformity. The Zoo Hypothesis is an example of a theory that violates the principle.

observation selection effect
The bias that occurs when the way an observer interprets evidence is filtered by the way the evidence was collected or by the fact that the observer exists to collect the evidence.

Order of the Dolphin
The tongue-in-cheek name adopted by attendees at the first SETI meeting in 1961 organized by Frank Drake and J. P. T. Pearman. The order included Drake, Pearman, Dana Atchley, Melvin Calvin, Shu-Shu Huang, John Lilly, Philip Morrison, Bernard Oliver, Carl Sagan, and Otto Struve.

Orion Arm
A minor spiral arm of the Milky Way. Our solar system is located on the inner edge of the Orion Arm, about halfway along its length, 26,000 light years from the galactic center.

plurality of worlds
The idea necessitated by an atomist, materialist, or Copernican point of view that Earth is not unique but must be one of many worlds, some possibly inhabited by other intelligent beings.

Project Cyclops
A proposed project, described in a NASA report in 1971, to build a large array of radio telescopes to search for signals from intelligent life at a cost of $6 billion to $10 billion.

Project Ozma
The first systematic attempt to listen for radio signals from extraterrestrial civilizations, carried out by radio astronomer Frank Drake in mid-1960 using the Tatel Telescope at the National Radio Astronomy Observatory in Green Bank, West Virginia.

radial-velocity method
A method for detecting exoplanets orbiting distant stars by measuring tiny wobbles in the stars' movements in the direction toward Earth or away from Earth.

Rare Earth Hypothesis
The suggestion, championed by University of Washington scientists Peter Ward and Donald Brownlee, that complex animal life is unique to Earth due to an unrepeatable combination of local circumstances.

RNA
Ribonucleic acid, a single-stranded chain of nucleotides that carry instructions from a cell's DNA to its ribosomes, where it is used to direct protein synthesis.

Sagittarius A*
A radio source at the center of the Milky Way galaxy that is thought to include a supermassive black hole.

SERENDIP
The Search for Extraterrestrial Radio Emissions from Nearby Developed Intelligent Populations: a long-running project of the Berkeley SETI Research Center.

SETI
The search for extraterrestrial intelligence: the practical and theoretical pursuit of evidence for intelligent life elsewhere in the universe. Now divided into "active SETI" (see CETI and METI) and "passive SETI," or conventional scanning for incoming radio or optical signals.

SETI Institute
A nonprofit research institute based in Mountain View, California, and set up by SETI scientist Jill Tarter in 1984 to study topics in astronomy, astrobiology, geoscience, and exoplanet research as well as to carry out passive SETI searches.

SETI@home
A long-running experiment at the SETI Research Center at the University of California–Berkeley to gather radio data and analyze it for signals of artificial extraterrestrial origin using a distributed network of Internet-connected computers owned by volunteer members.

shadow biosphere
The hypothesis that abiogenesis occurred more than once on Earth and that one or more ecosystems of non-DNA-based life may still exist.

transit method
A method for detecting exoplanets in which telescopes measure the slight dimming of a star's light caused when a planet "transits" the star—that is, passes between the star and the telescope.

Viking
The NASA mission to send two robotic landers to Mars in 1976. The Viking landers were equipped with elaborate biological experiment systems to test for the presence of organic activity in Martian soil.

Voyager
The NASA mission to send two robotic probes to the outer solar system on relatively rapid "grand-tour" trajectories that took them past Jupiter and Saturn as well as (in Voyager 2's case) Uranus and Neptune.

Voyager Interstellar Record
Twin gold-plated phonograph records mounted to the sides of the Voyager probes as time capsules and messages to hypothetical extraterrestrial civilizations who might intercept the craft.

water hole
The range of microwave frequencies inside the microwave window that includes the hydrogen line and the spectra of hydroxyl ions.

Zoo Hypothesis
The idea that extraterrestrial civilizations exist but are quarantining Earth until humans reach some requisite level of intellectual or social advancement.

NOTES

Preface

1. Paul Davies, *The Eerie Silence: Renewing the Search for Alien Intelligence* (Boston: Houghton Mifflin Harcourt, 2010), 2.

2. There is, however, a lively debate under way about whether 'Oumuamua, a large interstellar object that passed through the inner solar system in the fall of 2017, might have been a light sail or some other kind of craft built by an extraterrestrial civilization. See chapter 5 for more discussion of 'Oumuamua.

3. Carl Sagan, *The Cosmic Connection* (New York: Doubleday, 1973).

4. Quoted in Wade Roush, "Spielberg Finances E.T. Search," *Harvard Independent*, October 3, 1985.

5. Carl Sagan, *Contact* (New York: Random House, 1980).

Introduction

1. For the story of Eratosthenes's measurement, see, for example, Nicholas Nicasro, *Circumference: Eratosthenes and the Ancient Quest to Measure the Globe* (New York: St. Martin's Press, 2008).

2. For a more serious treatment of Native American culture in the pre-Columbian moment, see Charles Mann, *1491: New Revelations of America before Columbus* (New York: Knopf, 2005).

3. The account here comes mainly from the Harvard physicist and SETI scientist Paul Horowitz, who heard it from York. See "The Fermi Paradox," late 1998, http://seti.harvard.edu/unusual_stuff/unpublished/fermi.htm. But there are several other versions; in some, the conversation took place in 1943, not in 1950. See David Grinspoon, *Lonely Planets: The Natural Philosophy of Alien Life* (New York: Harper Collins, 2003), 311.

4. Milan M. Ćirković, *The Great Silence: The Science and Philosophy of Fermi's Paradox* (Oxford: Oxford University Press, 2018), 2.

5. Peter Ward and Donald Brownlee, *Rare Earth: Why Complex Life Is Uncommon in the Universe* (New York: Copernicus, 2000).

6. Stephen Webb, *If the Universe Is Teeming with Aliens ... Where Is Everybody? Seventy-Five Solutions to the Fermi Paradox and the Problem of Extraterrestrial Life*, 2nd ed. (New York: Springer, 2015).

7. NASA Exoplanet Archive, https://exoplanetarchive.ipac.caltech.edu/docs/counts_detail.html, data retrieved April 30, 2019.

8. Michael H. Hart, "An Explanation for the Absence of Extraterrestrials on Earth," *Quarterly Journal of the Royal Astronomical Society* 16 (1975): 128–135.

9. Quoted in Joel Achenbach, and Peter Essick, "Life beyond Earth," *National Geographic Magazine*, January 2000.

Chapter 1

1. Barbara Duncan, *The Origin of the Milky Way and Other Living Stories of the Cherokee* (Chapel Hill: University of North Carolina Press, 2008).

2. Karl Taube, *Aztec and Maya Myths* (Austin: University of Texas Press, 1993), 45–47.

3. Karl Popper, *Conjectures and Refutations: The Growth of Scientific Knowledge* (New York: Routledge, 2014), 186.

4. Richard McKirhan, "Anaximander's Infinite Worlds," in *Essays in Ancient Greek Philosophy VI: Before Plato*, ed. Anthony Preus (Albany: State University of New York Press, 2001), 49–66.

5. Bertrand Russell, "The Atomists," in *History of Western Philosophy: Collectors Edition* (New York: Routledge, 2009), chap. 9.

6. This version of the Metrodorus quote comes to us via Aëtius and the unknown authors grouped under the name "Pseudo-Plutarch." See Plutarch, *Morals*, chap. 5, https://ebooks.adelaide.edu.au/p/plutarch/nature/book1.html#chapter5.

7. Epicurus to Herodotus, in Epicurus, *The Epicurus Reader: Selected Writings and Testimony*, ed. Lloyd P. Gerson (Indianapolis, IN.: Hackett, 1994), 8.

8. Lucretius, *De rerum natura*, ed. William Ellery Leonard (1916), http://www.perseus.tufts.edu/hopper/text?doc=Perseus%3Atext%3A1999.02.0131%3Abook%3D2%3Acard%3D1048.

9. Plato, *The Timaeus*, trans. Benjamin Jowett (1892), http://classics.mit.edu/Plato/timaeus.html.

10. Aristotle, *On the Heavens*, trans. J. L. Stocks (Adelaide, Australia: University of Adelaide, 2016), book I, chap. 8, https://ebooks.adelaide.edu.au/a/aristotle/heavens/index.html.

11. William Whewell, *Of the Plurality of Worlds: An Essay, Also a Dialogue on the Same Subject* (London: Parker, 1855), 144.

12. Benjamin D. Wiker, "Alien Ideas: Christianity and the Search for Extraterrestrial Life," *Crisis*, November 4, 2002, https://web.archive.org/web/20030210140752/http://www.crisismagazine.com/november2002/feature7.htm.

13. Wiker, "Alien Ideas."

14. Stephen Greenblatt, *The Swerve: How the World Became Modern* (New York: Norton, 2011).

15. Johannes Kepler, *Kepler's Conversation with Galileo's Sidereal Messenger*, trans. Edward Rosen (New York: Johnson, 1965), 42.

16. Galileo Galilei, *Letters on Sunspots*, in *Discoveries and Opinions of Galileo*, trans. Stillman Drake (New York: Doubleday, 1957) 137.

17. One valuable book that does attempt such a comprehensive overview is Steven J. Dick, *Plurality of Worlds: The Origins of the Extraterrestrial Life Debate from Democritus to Kant* (Cambridge: Cambridge University Press, 1984). For another thorough tour of discussions of extraterrestrials in the seventeenth, eighteenth, and nineteenth centuries, see Michael J. Crowe, *The Extraterrestrial Life Debate, 1750–1900* (Cambridge: Cambridge University Press, 1986).

18. Bernard le Bovier Fontenelle, *Conversations on the Plurality of Worlds*, trans. H. A. Hargreaves. (Berkeley: University of California Press, 1990), 49, 60.

19. Christian Huygens, *The Celestial Worlds Discover'd; or, Conjectures Concerning the Inhabitants, Plants, and Productions of the Worlds in the Planets* (London: James Knapton, 1698) 149, 151.

20. Whewell's arguments are summarized in Crowe, *Extraterrestrial Life Debate*, chap. 6, sec. 3.

21. Whewell, *Of the Plurality of Worlds*, 330–331, emphasis added.

22. Giovanni Schiaparelli, *La via sul pianeta Marte: Tre scritti di Schiaparelli su Marte e i "marziani,"* ed. Pasquale Tucci, Agnese Mandrino, and Antonella Testa (Milan: Mimesis, 1998), 76; translation courtesy of Paola Rebusco.

23. Percival Lowell, *Mars* (Boston: Houghton, Mifflin, 1895), 149–150.

24. Lowell, *Mars*, 209.

25. A. R. Wallace, *Is Mars Habitable?* (London: Macmillan and Co., 1907), 38–77.

26. A hundred million of millions to one is 10^{14} to 1: small odds indeed. See the appendix to A. R. Wallace, *Man's Place in the Universe*, 4th ed. (London: Chapman and Hall, 1904).

27. Here I must thank Carl Sagan for introducing 13-year-old me to the story of Percival Lowell in "Blues for Red Planet," episode 5 of the television series *Cosmos*, PBS, October 26, 1980. In the book version, Sagan wrote: "Lowell always said that the regularity of the canals was an unmistakable sign that they were of intelligent origin. This is certainly true. The only resolved question was which side of the telescope the intelligence was on" (*Cosmos*, 110).

Chapter 2

1. Guiseppe Cocconi and Philip Morrison, "Searching for Interstellar Communications," *Nature*, September 19, 1959, 846, emphasis added.

2. Cocconi and Morrison, "Searching for Interstellar Communications," 845.

3. For the details of Drake's Project Ozma, see Grinspoon, *Lonely Planets*, 163; Sarah Scoles, *Making Contact: Jill Tarter and the Search for Extraterrestrial Intelligence* (Berkeley, CA: Pegasus Books, 2017), 60–64; and Davies, *The Eerie Silence*, 1.

4. Frank Drake and Dava Sobel, "The Origin of the Drake Equation," *Astronomy Beat* 46 (April 5, 2010): 1. Drake's statement that only 10 people in the world were thinking about extraterrestrial life in 1961 was a bit of an exaggeration. For a thorough look at the debate at that time, see Steven Dick, *The Biological Universe: The Twentieth-Century Extraterrestrial Life Debate and the Limits of Science* (Cambridge: Cambridge University Press, 1996).

5. Drake and Sobel, "The Origin of the Drake Equation," 2–3.

6. Drake and Sobel, "The Origin of the Drake Equation," 3.

7. David Grinspoon, *Earth in Human Hands: Shaping Our Planet's Future* (New York: Grand Central Publishing, 2016), 299–305.

8. L. M. Gindilis and L. I. Gurvits, "SETI in Russia, USSR, and the post-Soviet Space: A Century of Research," *Acta Astronautica* 162 (September 2019), https://doi.org/10.1016/j.actaastro.2019.04.030.

9. I. S. Shklovskii, and Carl Sagan, *Intelligent Life in the Universe* (San Francisco: Holden-Day, 1966), 359–360.

10. NASA, *Project Cyclops: A Design Study of a System for Detecting Extraterrestrial Intelligent Life*, NASA Report no. CR 11445 (Washington, DC: NASA, 1971), 1.

11. NASA, *Project Cyclops*, 4.

12. Scoles, *Making Contact,* 65.

13. Quoted in Grinspoon, *Earth in Human Hands*, 313.

14. Quoted in Bill Steele, "It's the 25th Anniversary of the First Attempt to Phone E.T.," *Cornell Chronicle*, November 12, 1999, http://news.cornell.edu/stories/1999/11/25th-anniversary-first-attempt-phone-et-0.

15. Quoted in Steven Johnson, "Greetings, E.T. (Please Don't Murder Us)," *New York Times Magazine*, June 28, 2017.

16. Quoted in Alan Penny, "The SETI Episode in the 1967 Discovery of Pulsars," *European Physical Journal*, February 2013, 6.

17. Robert Krulwich, "Aliens Found in Ohio? The 'Wow' Signal," *Weekend Edition Saturday*, National Public Radio, May 28, 2010, https://www.npr.org/sections/krulwich/2010/05/28/126510251/aliens-found-in-ohio-the-wow-signal.

18. Scoles, *Making Contact*, 67. Scoles's book was my main source for the details of Tarter's work.

19. Paul Horowitz, "A Search for Ultra-narrowband Signals of Extraterrestrial Origin," *Science* 201 (August 25, 1978): 733–735.

20. Roush, "Spielberg Finances E.T. Search."

21. Quoted in Zeeya Merali, "Search for Extraterrestrial Intelligence Gets a $100-Million Boost," *Nature*, July 20, 2015, 392–393.

22. NASA, *Project Cyclops*, 64.

23. Anita Heward, "LOFAR Opens Up the Low-Frequency Universe— and Starts a New SETI Search," *Phys.org*, April 14, 2010, https://phys.org/news/2010-04-lofar-low-frequency-universe-seti.html.

24. T. Joseph W. Lazio, Jill Tarter, and D. J. Wilner, "Cradle of Life," *Science with the Square Kilometer Array*, 2004, https://www.skatelescope.org/cradle-life.

25. Hillary Lebow, "Search for Extraterrestrial Intelligence Expands at Lick Observatory," UC Santa Cruz Newscenter, March 23, 2015, https://news.ucsc.edu/2015/03/lick-niroseti.html.

26. SETI Institute, Technosearch, https://technosearch.seti.org.

27. Quoted in SETI Institute, "New Search for Signals from 20,000 Star Systems Begins," press release, March 30, 2016, https://www.seti.org/seti-institute/press-release/new-search-signals-20000-star-systems-begins.

28. Breakthrough Initiatives, "National Astronomical Observatories of China, Breakthrough Initiatives Launch Global Collaboration in Search for Intelligent Life in the Universe," press release, October 12, 2016, http://astrobiology.com/2016/10/national-astronomical-observatories-of-china-breakthrough-initiatives-launch-global-collaboration-in.html.

29. Jason Daley, "In the Search for Aliens, We've Only Analyzed a Small Pool in the Cosmic Ocean," *Smithsonian*, October 2, 2018, https://www.smithsonianmag.com/smart-news/search-aliens-weve-only-examined-cosmic-hot-tub-180970447.

30. Jason T. Wright, Shubham Kanodia, and Emily Lubar, "How Much SETI Has Been Done? Finding Needles in the n-Dimensional Cosmic Haystack," Arxiv.org astro-ph, September 19, 2018, https://arxiv.org/abs/1809.07252.

Chapter 3

1. Davies, *The Eerie Silence*, 25.

2. Davies, *The Eerie Silence*, 32.

3. For the story of Carl Woese, see David Quammen, *The Tangled Tree: A Radical New History of Life* (New York: Simon and Schuster, 2018).

4. David Toomey, *Weird Life: The Search for Life That Is Very, Very Different from Our Own* (New York: Norton, 2013), 4–11.

5. Toomey, *Weird Life*, 28.

6. Douglas Fox, "Lakes under the Ice: Antarctica's Secret Garden," *Nature*, August 21, 2014, 244–246.

7. Catherine Offord, "Life Thrives within the Earth's Crust," *The Scientist*, October 2018, https://www.the-scientist.com/features/life-thrives-within-the-earths-crust-64805.

8. US National Research Council, *The Limits of Organic Life in Planetary Systems* (Washington, DC: National Academies Press, 2007), 31.

9. Leonard David, "NASA's Mars Rover Curiosity Had Planetary Protection Slip-up," *Scientific American*, December 1, 2011, https://www.scientificamerican.com/article/nasas-mars-rover-curiositt/, and Jyoti Madhusoodanan, "Microbial Stowaways to Mars Identified," *Nature*, May 19, 2014, https://www.nature.com/news/microbial-stowaways-to-mars-identified-1.15249.

10. Melissa Gaskill, "Space Station Research Shows That Hardy Little Travelers Could Colonize Mars," NASA Johnson Space Center news release, May 2, 2014, https://www.nasa.gov/mission_pages/station/research/news/eu_tef.

11. Using modern data-analysis software, Levin's allies say they have found evidence of circadian rhythms in the LR experiment's radiation measurements, another possible signal of life. See Ker Than, "Life on Mars Found by NASA's Viking Mission?" *National Geographic News*, April 15, 2012, https://news.nationalgeographic.com/news/2012/04/120413-nasa-viking-program-mars-life-space-science.

12. NASA, *Viking 40th Anniversary: Life on Mars*, EDGE video, https://www.nasa.gov/mission_pages/Viking.

13. Davies writes: "Gil wanted to run the LR experiment with two broths, one having left-handed amino acids and right-handed sugars, the other using their mirror forms. Thus, had the Mars soil fizzed equally for both, a simple chemical reaction would be the most likely explanation—the one most scientists now back. But if biology had been responsible, then there would have been a marked difference in response between the two forms of broth" (*The Eerie Silence*, 39).

14. Mike Wall, "Signs of Life on Europa May Be Just beneath the Surface," *Scientific American*, July 23, 2018, https://www.scientificamerican.com/article/signs-of-life-on-europa-may-be-just-beneath-the-surface.

15. US National Research Council, *Limits of Organic Life in Planetary Systems*, 30–31.

16. James Stevenson, Jonathan Lunine, and Paulette Clancy, "Membrane Alternatives in Worlds without Oxygen: Creation of an Azotosome," *Science Advances*, February 27, 2015, http://advances.sciencemag.org/content/1/1/e1400067.

17. See chapter 3, "A Shadow Biosphere?," in Davies, *The Eerie Silence*, 42–65.

18. US National Research Council, *Limits of Organic Life in Planetary Systems*, 74–75.

19. Donald Goldsmith's book *Exoplanets: Hidden Worlds and the Search for Extraterrestrial Life* (Cambridge, MA: Harvard University Press, 2018) is a wonderful source on the exoplanet story.

20. NASA Exoplanet Archive, https://exoplanetarchive.ipac.caltech.edu/, data retrieved July 27, 2019. The archive offers up-to-date information about the exoplanet hunt.

21. Seth Shostak, "This Weird Planetary System Seems Like Something from Science Fiction," *Mach*, *NBC News*, February 22, 2017, https://www.nbcnews.com/mach/space/weird-planetary-system-seems-something-science-fiction-n724136.

22. See the Habitable Exoplanets Catalog maintained by the Planetary Habitability Laboratory at the University of Puerto Rico at Arecibo, http://phl.upr.edu/projects/habitable-exoplanets-catalog.

23. Sara Seager, William Bains, and Janusz Jura Petkowski, "Toward a List of Molecules as Potential Biosignature Gases for the Search for Life on Exoplanets and Applications to Terrestrial Biochemistry," *Astrobiology* 16 (2016): 465.

24. US National Research Council, *Limits of Organic Life in Planetary Systems*, 84.

Chapter 4

1. I'm referring mainly to Milan Ćirković, who considers the Drake Equation to be not just outmoded but also dangerous: "In the SETI field, invocation of the Drake equation is nowadays largely an admission of failure … to develop a real theoretical grounding for the search" (*The Great Silence*, 95).

2. Matthew Cobb, "Alone in the Universe: The Improbability of Alien Civilisations," in *Aliens: The World's Leading Scientists on the Search for Extraterrestrial Life*, ed. Jim al-Khalili (New York: Picador, 2016), 166.

3. "On the Shores of the Cosmic Ocean," episode 1 of *Cosmos*, PBS, September 28, 1980.

4. NASA Exoplanet Archive, https://exoplanetarchive.ipac.caltech.edu/docs/counts_detail.html.

5. Webb, *If the Universe Is Teeming with Aliens*, 230–234.

6. Michael Hart, "Habitable Zones about Main Sequence Stars," *Icarus* 37, no. 1 (January 1979): 351–357.

7. Erik Petigura, Andrew Howard, and Geoffrey Marcy, "Prevalence of Earth-Size Planets Orbiting Sun-Like Stars," *Proceedings of the National Academy of Sciences* 110, no. 48 (November 26, 2018): 19273–19278.

8. Ward and Brownlee, *Rare Earth*, 190–220.

9. Webb, *If the Universe Is Teeming with Aliens*, 288–290. To be clear, although Ward and Brownlee were aware of the Mars hypothesis, it wasn't a big part of their argument.

10. Ward and Brownlee, *Rare Earth*, 243.

11. Ward and Brownlee, *Rare Earth*, 250.

12. US National Research Council, *Limits of Organic Life in Planetary Systems*, 1.

13. David J. Darling, *Life Everywhere: The Maverick Science of Astrobiology* (New York: Basic Books, 2001), 103.

14. Ćirković, *The Great Silence*, 152.

15. John G. Cramer, "The Pump of Evolution," *Analog Science Fiction & Fact*, January 1986, https://www.npl.washington.edu/av/altvw11.html.

16. Ćirković, *The Great Silence*, 172.

17. This is known as the Adaptationist or Permanence Hypothesis, after a story by science-fiction author Karl Schroeder. See Ćirković, *The Great Silence*, 158–162.

18. Ross Andersen, "What the Crow Knows," *Atlantic*, March 2019, https://www.theatlantic.com/magazine/archive/2019/03/what-the-crow-knows/580726.

19. Sagan, *Cosmos*, 301. Note that Sagan's version of the Drake Equation was slightly different from the standard one. He used N^* (the absolute number of stars in the galaxy) instead of R^* (the rate of star formation), and f_L ("the fraction of a planetary lifetime graced by civilization") instead of L. But the math comes out the same. Note also that Earth will become uninhabitable in about one billion years, long before the sun dies.

20. The Light-Cage Hypothesis: see Webb, *If the Universe is Teeming with Aliens*, 101–103.

21. The Galactic Stomach Ache Hypothesis: see Ćirković, *The Great Silence*, 222–228.

22. The Thoughtfood-Exhaustion Hypothesis: see Ćirković, *The Great Silence*, 163–164.

23. The Deadly Probes Hypothesis: see Ćirković, *The Great Silence*, 187–193.

24. The Astrobiological Phase Transition Hypothesis: see Ćirković, *The Great Silence*, 174–178.

25. Nick Bostrom, *Anthropic Bias: Observation Selection Effects in Science and Philosophy* (New York: Routledge, 2010).

26. Nick Bostrom, "Where Are They? Why I Hope the Search for Extraterrestrial Life Finds Nothing," *MIT Technology Review*, April 22, 2008, 120.

27. Robin Hanson, "The Great Filter—Are We Almost Past It?" September 15, 1998, http://mason.gmu.edu/~rhanson/greatfilter.html.

28. David Deutsch, *The Beginning of Infinity: Explanations That Transform the World* (New York: Penguin Books, 2011), 446.

29. Ćirković explores the Hermit Hypothesis and finds it wanting (*The Great Silence*, 27–30). It assumes that every individual in a hermit species feels the same way and that the species has figured out how to avoid leaking any transmissions or other information about themselves.

30. Webb, *If the Universe Is Teeming with Aliens*, 183–185.

31. The Sustainability or Aliens Are Green Hypothesis: see Ćirković, *The Great Silence*, 220–222, and Webb, *If the Universe Is Teeming with Aliens*, 106–109.

32. The Resource-Exhaustion Hypotheses: see Ćirković, *The Great Silence*, 185, and Webb, *If the Universe Is Teeming with Aliens*, 103–104.

33. Webb, *If the Universe Is Teeming with Aliens*, 111–113.

34. The Distance-Learners Hypothesis: see Webb, *If the Universe Is Teeming with Aliens*, 187–189.

35. Ćirković calls this the "Introvert Big Brother" Hypothesis: see *The Great Silence*, 182–185.

36. The Persistence Hypothesis, also known as the Percolation Hypothesis: see Ćirković, *The Great Silence*, 212–214, and Webb, *If the Universe Is Teeming with Aliens*, 92–98.

37. Jonathan Carroll-Nellenback, Adam Frank, Jason Wright, and Caleb Shaw, "The Fermi Paradox and the Aurora Effect: Exo-civilization Settlement, Expansion, and Steady States," ArXiv preprint, February 13, 2019, https://arxiv.org/pdf/1902.04450.pdf.

38. The average distance between any two communicating civilizations is calculated using a standard formula for the number of spheres of a given volume that fit into a space of a given volume. The formula is ((space-volume/sphere-volume)/packing-density), where the packing density is the optimal 0.74048 for cubical or hexagonal packing. We know the number of spheres, 16,875 in this case, and the volume of the galaxy, so we can solve for sphere volume and hence the sphere radius. The distance between any two communicative civilizations in this idealized scenario will be twice this radius.

39. "Kepler-1229b," *Wikipedia*, n.d., https://en.wikipedia.org/wiki/Kepler-1229b.

40. Douglas Adams, *The Hitchhiker's Guide to the Galaxy* (London: Pan Books, 1979), chap. 8.

41. Ćirković calls this the "Eternal Wanderers" Hypothesis: see *The Great Silence*, 214–220.

42. This is sometimes called the Berserker Hypothesis: see Grinspoon, *Earth in Human Hands*, 348–351, and Webb, *If the Universe Is Teeming with Aliens*, 122–123.

43. Mark Buchanan, "Searching for Trouble?" *Nature Physics*, August 2016, 720.

44. Quoted in Johnson, "Greetings, E.T. (Please Don't Murder Us)."

45. John A. Ball, "The Zoo Hypothesis," *Icarus* 19 (1973): 347–349.

46. The nonexclusivity principle is one of the most powerful ideas in Milan Ćirković's book *The Great Silence* (85–90).

47. Stephen Baxter, "The Planetarium Hypothesis—a Resolution of the Fermi Paradox," *Journal of the British Interplanetary Society*. 54 (2001): 210–216.

48. Jason Koebler, "Elon Musk Says There's a 'One in Billions' Chance That Reality Is Not a Simulation," *Motherboard*, June 2, 2016, https://motherboard.vice.com/en_us/article/8q854v/elon-musk-simulated-universe-hypothesis.

49. J. Richard Gott, "Implications of the Copernican Principle for Our Future Prospects," *Nature*, May 27, 1993, 315–319.

50. For more discussion of the Delta-T argument, see Webb, *If the Universe is Teeming with Aliens*, 178–183. For a recent book on Gott's idea, see William Poundstone, *The Doomsday Calculation: How an Equation That Predicts the Future Is Transforming Everything We Know about Life and the Universe* (Boston: Little, Brown Spark, 2019).

51. Webb, *If the Universe Is Teeming with Aliens*, 208–211.

52. The Transcension Hypothesis: see Ćirković, *The Great Silence*, 195–199, and Webb, *If the Universe Is Teeming with Aliens*, 196–198.

53. Ćirković, *The Great Silence*, 133.

Chapter 5

1. Thomas Levenson, the head of MIT's science-writing program, tells the Vulcan story in compelling detail in *The Hunt for Vulcan … and How Albert Einstein Destroyed a Planet, Discovered Relativity, and Deciphered the Universe* (New York: Random House, 2015).

2. Ćirković calls this the "Paranoid Style in Galactic Politics" Hypothesis; see *The Great Silence*, 124–126.

3. Webb, *If the Universe Is Teeming with Aliens*, 160.

4. Marek Abramowicz, *How to Search for a Signal from an Alien Civilization*, video, December 4, 2018, https://www.youtube.com/watch?v=P-XE7DOFLo0.

5. A double-size DVD-RAM disk holds 9.4 gigabytes of data. Assume that it flies for one second in a small room. You have just sent data at 9.4 gigabytes per second or 75,200 megabits per second. Compared to a maximum download speed for most home broadband services (circa 2020) of 300 megabits per second, the flying disk offers a 250-times improvement.

6. Webb, *If the Universe Is Teeming with Aliens*, 161–163.

7. Shmuel Bialy and Abraham Loeb, "Could Solar Radiation Pressure Explain 'Oumuamua's Peculiar Acceleration?" accepted for publication in *Astrophysical Journal Letters*, November 6, 2018; Abraham Loeb, "6 Strange Facts about the Interstellar Visitor 'Oumumua," *Scientific American*, November 20, 2018, https://blogs.scientificamerican.com/observations/6-strange-facts -about-the-interstellar-visitor-oumuamua.

8. Quoted in Josh Swartz, "Harvard Astronomer on Why Aliens Aren't Science Fiction," WBUR, January 30, 2019, https://www.wbur.org/ endlessthread/2019/01/30/oumuamua-alien-probe-avi-loeb.

9. Andreas Hein, Nikolaos Perakis, T. Marshall Eubanks, Adam Hibberd, Adam Crowl, Kieran Hayward, Robert G. Kennedy III, et al., "Project Lyra: Sending a Spacecraft to 1I/'Oumuamua (Former A/2017 U1), the Interstellar Asteroid," ArXiv.org, October 19, 2018, https://arxiv.org/ftp/arxiv/papers/1711/ 1711.03155.pdf.

10. Quoted in Oded Carmeli, "If True, This Could Be One of the Greatest Discoveries in Human History," *Haaretz*, January 16, 2019, https://www.haaretz .com/us-news/.premium.MAGAZINE-if-true-this-could-be-one-of-the-greatest -discoveries-in-human-history-1.6828318.

11. Nathalie Cabrol, "Alien Mindscapes—a Perspective on the Search for Extraterrestrial Intelligence," *Astrobiology* 16, no. 9 (2016): 663, 667.

12. Cabrol, "Alien Mindscapes," 669.

13. Four of my favorites films that focus on postcontact outcomes include *2001: A Space Odyssey* (1968), *Close Encounters of the Third Kind* (1977), *Contact* (1997), and *Arrival* (2016).

14. See Carl Jung, *Flying Saucers: A Modern Myth of Things Seen in the Skies* (Princeton, NJ: Princeton University Press, 1979).

FURTHER READING

Billings, Lee. *Five Billion Years of Solitude: The Search for Life among the Stars*. New York: Penguin, 2013.

Ćirković, Milan. *The Great Silence: The Science and Philosophy of Fermi's Paradox*. Oxford: Oxford University Press, 2018.

Davies, Paul. *The Eerie Silence: Renewing the Search for Alien Intelligence*. Boston: Houghton Mifflin Harcourt, 2010.

Deutsch, David. *The Beginning of Infinity: Explanations That Transform the World*. New York: Penguin, 2011.

Goldsmith, Donald. *Exoplanets: Hidden Worlds and the Quest for Extraterrestrial Life*. Cambridge, MA: Harvard University Press, 2018.

Grinspoon, David. *Earth in Human Hands: Shaping Our Planet's Future*. New York: Grand Central Publishing, 2016.

Grinspoon, David. *Lonely Planets: The Natural Philosophy of Alien Life*. New York: Harper Collins, 2003.

Al-Khalili, Jim, ed. *Aliens: The World's Leading Scientists on the Search for Extraterrestrial Life*. New York: Picador, 2016.

Scoles, Sarah. *Making Contact: Jill Tarter and the Search for Extraterrestrial Intelligence*. Berkeley, CA: Pegasus Books, 2017.

Shostak, Seth. *Confessions of an Alien Hunter: A Scientist's Search for Extraterrestrial Intelligence*. New York: Penguin Random House, 2009.

Toomey, David. *Weird Life: The Search for Life That Is Very, Very Different from Our Own* New York: Norton, 2013.

Ward, Peter D., and Donald Brownlee. *Rare Earth: Why Complex Life Is Uncommon in the Universe*. New York: Copernicus, 2000.

Webb, Stephen. *If the Universe Is Teeming with Aliens ... Where Is Everybody? Seventy-Five Solutions to the Fermi Paradox and the Problem of Extraterrestrial Life*. 2nd ed. New York: Springer, 2015.

INDEX

The MIT Press Essential Knowledge Series

WADE ROUSH, a Boston-based science and technology journalist, is a columnist at *Scientific American* and the producer and host of *Soonish*, an independent podcast about the future. He has served as Boston bureau reporter for *Science*, senior editor and San Francisco bureau chief at *MIT Technology Review*, chief correspondent and San Francisco editor for *Xconomy*, and acting director of MIT's Knight Science Journalism program. He holds a PhD in the history and social study of science and technology from MIT.